D0256467

MOUNTAIN WEATHER

View from Rannerdale Knotts over Buttermere.
Cumulus and strato cumulus waves.
Photo: Berenice Pedgley

MOUNTAIN WEATHER

A PRACTICAL GUIDE FOR HILLWALKERS AND CLIMBERS IN THE BRITISH ISLES

by

David Pedgley

© David Pedgley 1997
ISBN 1 85284 256 3
Reprinted 1983, 1985, 1987, 1991
Revised 1994
Second Edition 1997
Reprinted 2000, 2004

HLP
ea
< P >
copy 1

Acknowledgements

The first edition of this book sprang from lectures given to outdoor activities instructors and courses for Mountain Leadership Certificates, as well as a series of week-long field courses on mountain weather held each summer in North Wales and sponsored by the Royal Meteorological Society. I am grateful to the participants and fellow instructors on those courses for their enthusiastic interest in mountain weather and for helping me to see the kind of book that is needed - not only a field guide for all who go into the mountains but also a reference for instructors in mountain outdoor activities generally. I thank the former Director of the National Mountain Centre, Plas-y-Brenin, and the principals of Glenmore Lodge and Benmore Centre for Outdoor Pursuits for their encouragement and help.

Results from these courses, along with those of other studies described in many technical reports and journals, were used to provide illustrations of the wide range of effects that mountains have on the weather. Since then, there has been much new work on mountain weather, sparked off not only by concern over lessening the spread of air pollution and the damage to forests and structures by strong winds but also by desire to improve forecasts. As by-products, much has been learnt that is of value to hill-walkers and climbers. Hence the need for this second edition. I thank the many who have been involved in the new work for the chance to use their results.

For the records from the automatic weather station on Cairn Gorm summit, I thank Heriot-Watt University, Edinburgh; for the satellite images, I thank the NERC Satellite Station, University of Dundee; and for the many maps and diagrams I thank my wife, Berenice, for transforming my rough sketches. The cloud and snowdrift photographs are my own.

David Pedgley

LANCASTER UNIVERSITY
30 NOV 2004
LIBRARY

Front Cover: Three kinds of clouds.
Flowerdale Forest seen from Beinn Alligin.
Photo: Andrew Pedgley

CONTENTS

INTRODUCTION

Our British mountains have plenty of weather. They are almost always windier, colder, cloudier and wetter than low country. What is more, the weather mood can change bewilderingly quickly – from gloomy, lowering skies and driving rain to shafts of sunlight and breath-taking patterns of colour. The weather can delight or endanger. It can create a mental picture never to be forgotten or thrust the unwary into a fight for life.

Whether it is to avoid danger or to add to enjoyment of the mountain scene, before starting a day outdoors among the hills it is wise to know what the weather is likely to be. This book helps hill-walkers and climbers to get the weather forecasts most suited to their needs, to understand them, and to modify them in the light of experience of mountain weather.

The book is in three parts.

PART ONE	describes the kinds of forecasts available and how they can be found
PART TWO	helps you understand the forecast through using weather maps
PART THREE	describes and explains some of the ways our mountains make their own weather.

All hill-walkers and climbers should read Part One, but if you want to understand the forecast, rather than just know it, Part Two will help you. It is not meant to be more than a start, for an understanding needs a knowledge of why the weather changes. That cannot be learned from a few pages; it comes from experience using weather maps, your eye, and further reading about the weather. (Some sources are listed at the end of this book.) The hardest bit is understanding how the mountains make their own weather. It is often said that mountain weather is unpredictable. That is not true, in principle, because it follows the same physical laws as the weather elsewhere. But, in practice, those laws can be difficult to apply to given places and times. You need plenty of experience. Part Three shows you some of the things to look for on a day among the hills. As your experience grows you will become better at modifying the forecast to suit your particular hills and valleys, for no forecast can yet do justice to all the peculiarities of mountain weather.

PART ONE
Weather Forecasts

A *forecast* states what the weather is *likely to be* at a given time and place. A *report* states what the weather *was* at a given time and place. The two are often confused.

There are many kinds of weather forecast. Use one prepared by a professional forecaster. It is unlikely to be right in every detail, but if you think you can do better test yourself daily for a few weeks - that should be enough to show you don't often do a better job. Remember, too, that a forecast can become out of date in only a few hours. If you have any doubt, check the latest forecast.

1. General Forecasts
National and regional forecasts, usually for one or two days ahead and giving general guidance on what the weather is likely to be over the British Isles, are broadcast on radio and television, printed in newspapers and available by telephone or fax. For example, TGO Weathercheck gives forecasts for four days, followed by an outlook for a further three days.

Phone 09001 112 plus:

West and North-west Scotland	235	Northern Ireland	238
North-east and East Scotland	236	Wales	242
North-west England	240		

For other services, consult local telephone directories.

It is also possible to get up-to-the -minute reports and forecasts on your PC by means of MIST (Meteorological Information Self-briefing Terminal). Subscribers are provided with maps of recent weather, reports from particular places, animated satellite and radar images, maps of forecast weather and much more. For details, contact the Met Office, London Road, Bracknell, Berks RG12 2SZ (tel: 01344 420242).

Radio and Television
Times of broadcasts can be obtained from the *Radio Times, TV Times* and newspapers. TV forecasts usually have satellite images of clouds and radar

images of rain, sometimes in animated sequences that provide an impressive indication of how weather systems develop. These forecasts add considerable detail to the *report* of weather in the recent past, but remember that detail can change rapidly even though the general pattern changes more slowly.

Newspapers

Most papers print forecasts, provided by either by the Met Office or by commercial services, but sometimes all too briefly. Some give reports of yesterday's weather from a list of places in Britain, and some have weather maps. Mostly, these maps are forecasts for some time on the day of issue, often mid-day; they should be carefully distinguished from other maps showing weather at some time during the previous day. Remember, too, that newspaper forecasts are not as up-to-date as those on radio and TV because of the time needed for papers to be printed and distributed.

2. Forecasts for hill-walkers and climbers

Local weather forecasts for mountainous areas of Britain are available on automatic telephone services. For example, the Met Office provides Mountaincall.

Phone 0891-500 plus:

Scotland Highlands West	441	Scotland Highlands East	442
Snowdonia	449	Peak District	433
Yorkshire Dales	748	Lake District	483

These include details of wind, cloud base, visibility, temperature and freezing level for two days, and a more general outlook for the area over the following few days. For other services, consult local telephone directories.

3. The Internet

Various internet sites provide information about recent weather as well as forecasts. Because the number and contents of these sites change over time, it is advisable to check what is currently available by first using search engines for 'mountain weather'. The following can be used to check what has been happening to the weather and how it might change.

Recent Weather

129.13.102.67/wz/pics/Rgbsyn.html

Weather map for UK (latest in 6-hourly sequence, with individual observations)

www.bbc.co.uk/weather/satellite.shtml

Satellite images for Europe (animated 3-hourly sequence for previous 24 hours - wait to download)

www.sat.dundee.ac.uk/auth.html

Satellite image for UK (more detail, but less frequent)

www.bbc.co.uk/weather/uk radar 6hr.shtml

Radar images for UK (animated hourly sequence for previous 6 hours - wait to download)

129.13.102.67/wz/pics/Rsfloc2.html

Lightning over Europe (accumulated hourly records for the day)

www.phy.hw.ac.uk/resrev/aws/awsgraph.htm

Cairn Gorm summit - latest temperatures and winds

www.fhc.co.uk/weather/current/cross.asp

Snowdon summit - latest temperature and winds

General Forecasts

For various regions

www.meto.govt.uk/sec3/flatfree.html

Animated 3-day sequence of weather maps for Europe (issued by Met Office)

www.bbc.co.uk/weather

UK, for day of issue (by BBC Weather Centre)

www.ananova.com/weather

UK, for day of issue (by Press Association Weather Centre)

Forecasts for Hill-walkers and Climbers

At present, only the following is available

weather.yahoo.com./forecast/Cairn Gorm UK c.html

Cairn Gorm summit, maximum and minimum temperatures for next 5 days

---

Know the forecast.
Understand it.
Be aware of how the hills
can change the weather.

PART TWO
Weather Maps

A weather map is a symbolic picture of the weather over a large area at a given time. Newspaper weather maps and those on TV are much simpler than those used by forecasters, but they show the temperature, wind and weather, say, over the British Isles or the whole of Europe. They can be used to help understand the forecast, for they give some idea of how the weather changes at a given place. Much of our weather is imported: the map shows which direction it is coming from and how long it will take to reach us.

The next 16 Sections give examples of weather maps for days with a wide range of weathers. Of course, the variations are endless. The usefulness of weather maps grows with experience, so try to understand each forecast in the light of the latest map and its likely changes with time.

Weather maps in this book have been drawn like those in newspapers. Each map of Britain shows the weather at 18 places for a particular time, and on some of them records from the automatic weather station on Cairn Gorm summit have been added, where available (as insets on the left to make them stand out). Temperature is given in degrees Centigrade (Celsius); wind speed in miles an hour is given inside the circles; wind direction is shown by an arrow head flying with the wind; and the weather is shown by the following letters.

b	sky largely or wholly blue		
bc	sky about half blue, half cloudy		
c	sky largely or wholly cloudy		
d	drizzle		
f	fog		
h	hail	**r**	rain
m	mist	**s**	snow
p	passing shower	**z**	haze

A SCALE FOR RATING WIND STRENGTH
(BASED ON THE BEAUFORT SCALE)

Force	Name used in forecasts	Speed range mph (kph)	You	Effects on		
				Surroundings	Lake surface	Fresh snow
0	calm	below 1 (below 2)	none	smoke rises straight up	like a mirror	none
1	light	1-3 (2-5)	none	smoke drifts	ripples	none
2		4-7 (6-11)	felt on face	grass and bracken quiver	wavelets but none breaking	none
3		8-12 (12-20)	hair ruffled, loose clothing flaps	heather and small twigs move	some wavelets breaking	a little drift near surface
4	moderate	13-18 (20-29)	hair disarranged	small branches move; loose dry grass picked up	some white horses	drifting up to a metre or so
5	fresh	19-24 (30-39)	walking inconvenienced	small trees swaying	many white horses	widespread drifting
6	strong	25-31 (40-50)	steady walking difficult	large branches move	some spray	some blowing above head height
7		32-38 (51-62)	walking with great difficulty	whole trees move	much spray	blowing in clouds above head height
8	gale	39-46 (63-74)	walking dangerous	twigs breaking from trees	foam in steaks along wind	dense blowing clouds
9		47-54 (75-87)	blown over crawling difficult	branches breaking from trees		
10	storm	55-63 (88-102)	progress impossible, even by crawling	some trees uprooted		
11		64-72 (103-116)		many trees uprooted		
12	hurricane	over 72 (over 116)		great damage		

1: Windy Day

Strong winds are tiresome; they slow us down and make us feel we are fighting a never-ending battle. Wind strength can be rated on a simple scale showing how much it affects us or our surroundings, such as vegetation, waves on a lake or snow being blown along the ground. The table opposite shows the scale used in some weather forecasts. It is hard to read a map in strong winds, and it is impossible to keep your eyes open continuously in a gale. Walking becomes very difficult: it may help to walk backwards, using a hand as a face guard; or better perhaps to wait in a sheltered spot for a lighter wind. You can be blown over in winds stronger than about 50mph.

Strong winds often last for hours, during which time the air can move hundreds of miles. Hence strong winds often cover large areas. These areas can be seen on weather maps - they are where the *isobars are close together*. An isobar is a line drawn on the weather map joining places where the weight of the atmosphere (its pressure) is the same. Weather maps show that isobars form patterns somewhat like contours on a topographic map. These patterns change from day to day. On a weather map each isobar corresponds to a particular pressure corrected to sea level. Pressure at sea level is always near 1,000 millibars (mbar) - over Britain it is rarely less than 950mbar or more than 1,050mbar. These differences seem small but they are very important because they tell us the strength and direction of the wind. On map 1A, lowest pressure (about 980mbar) is over northern Scotland; pressure increases outwards in all directions from this centre. Such a centre of low pressure is known as a *depression* or a *low*. The isobars on this day are closest over northern England - hence that is where the strongest winds are blowing. See also other maps in this book for examples of strongest winds where isobars are closest.

We often want to know not only the wind's strength but also its direction. We say the wind is 'west' or 'westerly' when it blows from west to east. Wind direction is that *from* which it is blowing. Winds blow more or less along the isobars such that if you stand with your back to the wind low pressure is on the left and high on the right

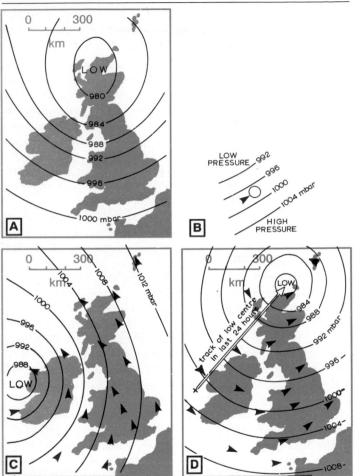

1A. Pattern of isobars showing a low, or depression, centred over northern Scotland.

1B. Wind direction related to run of isobars.

1C. A low centred just west of Ireland.

1D. The same low as C but a day later.

(diagram 1B). On map 1A there is therefore a south-west wind along the east coast of England and a north-west wind over Ireland. Look at any weather map to test these rules relating isobar direction and spacing to wind direction and speed.

The wind changes from day to day because the isobar patterns change. Lows and other patterns move (often at 20-40mph, or 1,000-2,000km a day), grow and later fade away over spells of a few days, only to be replaced by others.

If you look at a series of weather maps for Europe and the North Atlantic for a week, say in the newspapers, you will get some idea of how the isobar patterns change. Map 1C shows a low just west of Ireland, but one day later (map 1D) its centre has moved to just north of Scotland - it has moved about 1,000km in 24 hours at about 20mph. On map 1C the wind over Ireland was south, but on map 1D it has changed to north-west. A wind change in this sense - sunwise or clockwise - is a *veer*; in the other way it is a *backing*. Where are the strongest winds on map 1C?

Most lows that affect Britain come from the south-west, usually passing between Scotland and Iceland. Some take more southerly tracks, across Britain or even passing by to the south. Others track in from the west or north-west, and even from the south; and a few come at us from the east. In all cases, however, the changes in wind direction can be judged with the aid of a sequence of weather maps. Lows are more intense in winter, on average, but deep lows and gales can occur at any time of the year.

We know that the wind is usually gusty. Each gust reflects the passing of an eddy. Just as a river is full of ever-changing eddies with a great range of sizes, so is the atmosphere. The strongest gusts have speeds about half as much again as the average wind. So a force 6 wind (speed range 25-31mph) can be expected to have gusts up to 40mph (force 8). Likewise, a strong gale of 50mph will have gusts reaching 75mph (hurricane force). Gusts are more troublesome than the average wind because they are so unpredictable.

To be forewarned of likely strong winds to come, watch the weather maps for close isobars spreading over you, and don't forget the gusts.

2. Calm Day

Light winds are not always a boon. They may be good for sunbathing or fishing, but they don't help you keep cool when climbing.

Weather maps show where there are light winds - in several kinds of isobar patterns.

1. Map 2A shows an old low centred over England. It started several days ago but now it is filling - its pressure is rising. It is slow-moving and weak. The isobars are well spaced so the winds are light. Old, filling lows often have light winds over large areas.

2. Map 2B shows a centre of high pressure, known as an *anticyclone* or *high*. In its middle the pressure is almost uniform, so there are few isobars and they are far apart. A *high* often has light winds

2A. Weather map for Europe and the nearby Atlantic Ocean, showing a low over Britain.

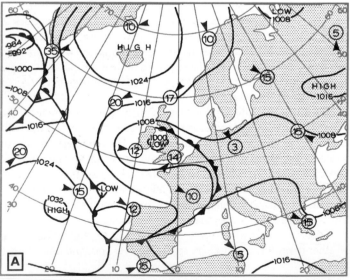

over large areas, and because it tends to be slow-moving the light winds can last for a few days.

3. Map 2C shows two double-centred lows and two highs. There is a *col* over England and Wales - where pressure is almost uniform, and so where there are few isobars and winds are light. The wind is south-west over northern Scotland, but east over southern England.

If you want little wind, watch the weather maps for an old slow-moving low, a high, a col, or anywhere with widely spaced isobars (map 14B is an example).

Even a good breeze by day can fall off at night to near calm, the more so well inland when there is little cloud about. Keep an eye open for widely spaced isobars and a cloudless sky (Section 7).

2B.   Weather map showing a high over Britain.

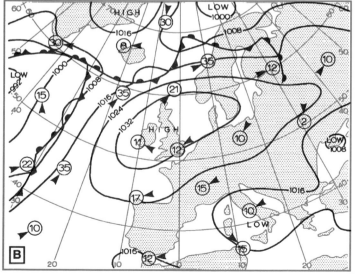

17

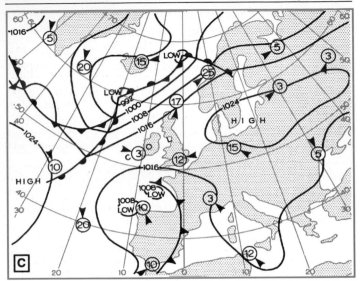

2C. Weather map showing a col over Britain.

3. Cold Day

We often think north winds are cold and south winds warm. This is broadly true, for north winds often come from colder latitudes nearer the pole. On map 3A there is a deep low over the North Sea, and north winds have spread to most of Britain. This map, and those of 3B, 6C, 7B, 15A and 31B show cold winds coming from high latitudes, and maps 15C, 15H and 16F show cold east winds. Sometimes the wind turns as it comes towards Britain. For example, map 3B shows a pattern of isobars known as a *trough* of low pressure; it lies south-

3A. A deep low over the North Sea, with strong northerly winds over Britain. A cold front with rain is clearing southern England, and wintry showers of hail, sleet and snow have spread to Scotland and northern Ireland.

3B. A trough of low pressure lies south-westwards across Britain from a low to the north-east. North winds over Scotland turn through north-west to west over southern England. Showers of rain and hail are widespread.

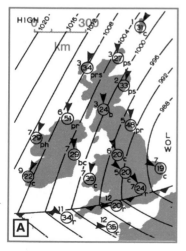

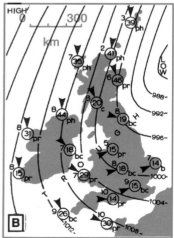

19

westwards from a low over Scandinavia. If the trough is slow-moving, cold north winds over Scotland turn to west as they reach England. Another example is map 3C, showing a pattern of isobars known as a *ridge* of high pressure; it lies south-eastwards from a high to the north-west of Scotland. North to north-east winds over the North Sea turn to east over England and Wales. Cold north winds can be found a day or two later having turned round to south - maps 3D and 15I are examples.

A weather map gives some idea of where the wind has come from and how quickly it has come. But because pressure patterns change it is often not easy to judge where the air has been in the last few days.

Sometimes warm and cold airstreams lie next to each other. The change-over is called a *front*, for it marks the leading edge where one

3C. A ridge of high pressure lies south-eastwards across Britain from a high to the north-west. North-west winds in the north turn to east over England, Wales and Ireland. Cloudy, but no rain.

3D. A low to the north-west and a warm front moving east across Ireland with rain and snow ahead. Cold south winds are being followed by mild south-westerlies.

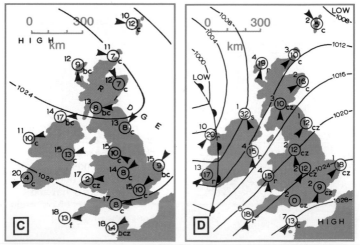

airstream replaces another. There are two main kinds of front. Map 3D shows a *warm front*; it is a line along which warm winds (here from the south-west, having reached western Ireland but will later spread to much of Britain) replace cold winds (here from the south). As a warm front passes overhead the wind veers and warm air starts to reach you. Map 3E shows a *cold front*; it is a line along which cold winds (here westerly and having already spread across Scotland, northern England and Wales) replace warm winds (here south-south-westerly over the rest of England). As a cold front passes overhead the wind veers and cold air starts to reach you.

Deep lows, with their strong winds, are caused by large temperature differences between high and low latitudes. Their fronts therefore tend to be strong, with large temperature changes (as in maps 3A, 3D and 3E).

Keep an eye on weather maps for fronts moving towards you. They will give a clue to likely changes to warmer or colder winds.

We can feel cold not only when the air is cold but also when the wind is strong, or we are inactive, or poorly clothed, or wet with rain, or in the shade. In weather forecasts, however, it is the air temperature that is given. A combination of low air temperature and strong wind

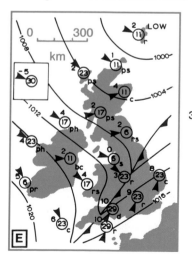

3E. A cold front moving south-east across Britain, with rain and drizzle turning to sleet and snow, and marking a change from mild south-south-west winds to cold north-westerlies bringing wintry showers of hail and snow to Ireland and Scotland. (Inset shows Cairn Gorm summit.)

can make you feel much colder than the air temperature would suggest. This *wind-chill* is based on a heat balance: between the rate generated by the body (varying with walking speed) and the rate lost to the wind (varying with wind speed, body size and shape, breathing rate, and insulating power of clothing). Wind-chill can be expressed as a temperature that would give a balance if only the wind were calm, at the same time assuming a fixed, standard set of controls on heat loss. So, for a given body size, etc, the wind-chill temperature gets lower than the air temperature as the wind speed is greater. But the great range of body size, etc., means that the standard set seldom occurs and a calculated wind-chill temperature is unlikely to apply to you. Remember, severe wind-chill can occur at any time of the year on British mountains. However, a wind-chill temperature below 0°C cannot cause freezing if the air temperature is above 0°C.

Beware the combination of low temperature and strong wind, the more so if there is rain or snow and you are poorly clothed.

4. Warm Day

We can feel warm not only when the air is warm but also when the wind is light, or we are too active, or too heavily clothed, or in the sun.

Warm winds often come to us from the south - map 4A is an example, and 4B is a satellite picture for the same time showing England and Wales cloudfree. The weather map may also show warm winds coming from other directions and turning as they approach Britain. For example, map 4C shows a winter high to the south-west, with mild north-west winds from the Atlantic covering Britain. The weather need not be cold despite north winds, the more so when clouds are broken and there is some sun to give added warmth.

Keep an eye on the weather map for winds coming from low latitudes. Watch for a warm front passing overhead - it will be followed by warm but often damp air. Watch for a slow-moving high to the east with light south winds and little cloud.

Broken clouds make a difference to the temperature, the more so

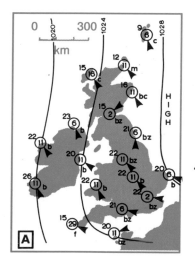

4A. A summer high over the North Sea, with warm and sunny south-east winds over Britain. The north-east coast is cool, where winds are off the sea, and there is sea fog around Cornwall.

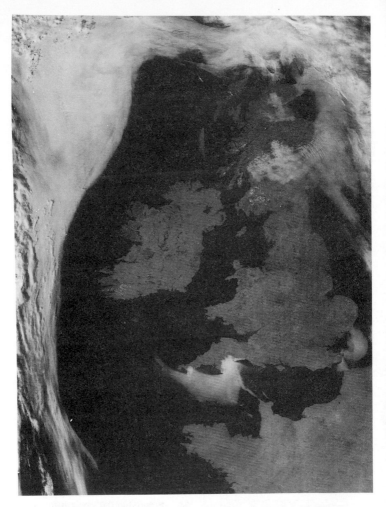

4B. Satellite picture for about the time of map 4A. England and Wales are cloud-free, but sea fog touches the coasts of Cornwall, Devon and Kent. An area of sea fog as large as Ireland lies to the north-west. Cumulus clouds have formed over the mountains of south-west and west Scotland and the Outer Hebrides, but not over the sea.

in summer, because sunshine heats the ground, which in turn heats the air. (Air lets most of the sunshine through; it is not warmed much directly.) That's why days with broken or little cloud are warmer than the nights; whereas days with deep, dark clouds are little warmer than the nights, the more so in winter (diagram 4D).

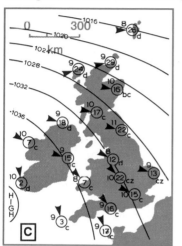

4C. A winter high to the south-west bringing mild but cloudy north-west winds and drizzle to exposed coasts and hills.

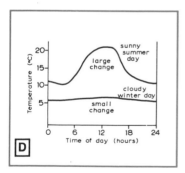

4D. Typical temperature changes during the day, contrasting a sunny summer day with a cloudy winter one.

5. Dull Day

Dull, sunless weather can be depressing, but for a long, active day outdoors it is often more bearable than endless sun. Dull skies can last for hours or days on end. If, at the same time, winds are strong, this means the cloud sheets are hundreds of kilometres across, if not thousands. These vast sheets of cloud show up very well on satellite pictures - see the examples in this book.

Almost all clouds are a result of the lifting of moist air. Air always contains some moisture in the form of invisible water vapour coming from the earth's surface - up to 3%. When the air is lifted the weight of the atmosphere on it gets less, so its expands and in so doing it cools (as air cools when rushing from the neck of an open balloon). If rising air cools enough a cloud of minute water droplets forms - typically 50-500 droplets in each cubic centimetre. Although a cloud may look 'solid' in fact it has typically only about 0.1% water by weight. Because a cloud follows the air motion its shape shows up the shape of the

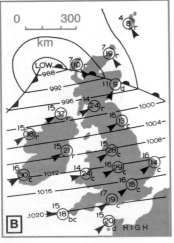

5B. Typical weather map when much stratus cloud is brought to Britain on moist south-west winds. Cool south-easterlies still over northern Scotland ahead of the warm front. Much rain and drizzle around the low.

5C. Tumbling eddies lift moist air to form stratus cloud.

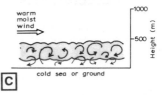

5D. Satellite picture showing widespread stratus cloud and sea fog over the Western Approaches. A tongue flows up the Irish Sea, where the eastern edge (from St David's Head across Cardigan Bay) shows the wind direction. Along the south coast of Ireland the cloud is just deep enough to give light drizzle before daytime heating changes it to cumulus inland.

rising air.

There are three main kinds of cloud that give a dull day.

1. *Stratus* clouds grow in a moist wind blowing over cold land or sea. Winds from low latitudes, or in an area of widespread rain, are very moist and can bring much low stratus cloud (colour photograph 5A), often with a base below 300m or even 100m. The air is lifted within countless tumbling eddies caused by the wind blowing over the rough surface of the earth (map 5B and diagram 5C). (See also map 3E for a moist airstream full of low stratus clouds; and maps 3D and 14C for examples when this cloud forms in long-lasting rain.) Satellite picture 5D shows widespread stratus clouds and sea fog over the Western Approaches and spreading up the Irish Sea. Over land, stratus clouds can clear away if the sun heats the ground enough, or if drier air spreads in to replace the moist, cloudy air (usually behind a front; see maps 3A and 3E for examples).

2. *Stratocumulus* clouds grow in a moist wind being gently lifted as a whole whilst it streams towards a low or a front (colour photograph 6H). Winds from the sea in highs and ridges can bring much stratocumulus cloud with a base often between 500 and 2,000m (see the south-west winds in map 5E and diagram 5F). The clouds

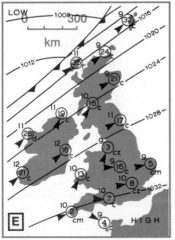

5E. Typical weather map when much stratocumulus cloud is brought to Britain in moist south-west winds ahead of a cold front approaching north-west Scotland. Some drizzle on exposed coasts and hills.

5F. Sheets of stratocumulus cloud form in air gently rising over a wide area.

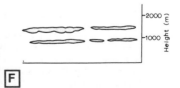

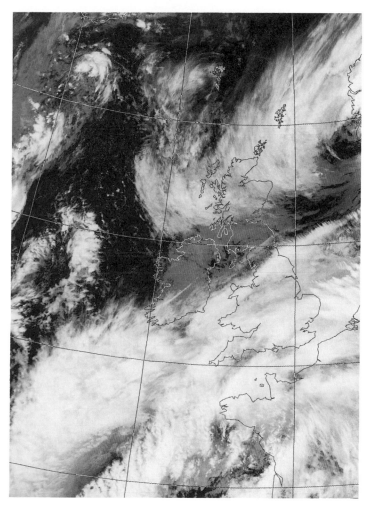

51. Satellite picture of the cloud sheet over England and Wales ahead of
the northward-moving warm front of the infamous storm of 1987 over
south-east England. Scotland has another cloud mass. (This is an
infra-red image, using heat radiated from cloud tops, not reflected
sunlight. Higher, and therefore colder, clouds are whiter. The white
clouds over England, Wales and northern Scotland are higher than
the grey clouds over southern Scotland and much of Ireland. The sea
is dark because it is warm.)

29

5J. Satellite picture for about the time of map 3E. Ireland and western Scotland can be seen beyond the marked western edge of the cloud band. Showers over western Scotland and western Ireland are not related to the mountains - they have come on the wind from over the sea.

(This is a night-time, infra-red picture.)

can clear away with a change of airstream, but sometimes a gentle, widespread sinking compresses, and therefore warms, the cloud enough to evaporate the droplets.

3. *Upper clouds* grow within some hundreds of kilometres of fronts and lows when the air at heights above about 3,000m is gently lifted over wide areas. They lie in vast bands along the fronts, which therefore bring spells of dull weather lasting many hours, or even a few days if they are slow-moving.

The clouds clear away as the front moves away, or as it weakens and pressure rises around it. The highest, with bases between about 6,000 and 12,000m, are thin and icy *cirrus* (if they are in tufts or streamers - see colour photograph 6A) or *cirrostratus* (if in milky sheets). Deeper, denser, greyer ice clouds, with the sun shining dimly as through ground glass, are *altostratus* (colour photograph 5G) with bases between about 3,000 and 5,000m, whereas sheets of water droplet clouds like stratocumulus are *altocumulus* (colour photograph 5H).

Upper cloud sheets in active lows and fronts are well organised, as shown in satellite pictures; but weak fronts and old, filling lows may have few upper clouds. Satellite picture 5I shows the vast afternoon cloud sheet over England and Wales ahead of the northward-moving warm front of the infamous 1987 storm, but Scotland has its own cloud mass. Picture 5J shows clouds on the cold front of map 3E: there is a clear western edge, beyond which Ireland and western Scotland are visible.

6. Cloudy Day

Cloudy days with only fitful sun are common. Broken clouds drift or rush across the sky, patchily shading the countryside. Such clouds are often *cumulus*, well known for their flat bases and rounded tops (colour photograph 6A). They grow when blobs or columns of warm air rise buoyantly from the sun-heated ground or warm sea (diagram 6B). Over land, they are mostly a daytime kind of cloud, starting in the morning as the ground warms up, and dying away around sunset as the ground cools down again. Over warm sea and windward coasts they can be seen both by day and by night when cold winds blow. North winds on the western side of a low are places where they are often found (map 6C, also 3A and 3B). Cloud base is often between 600 and 1,200m, higher for drier winds. When the sea is too cool to set off cumulus clouds, the daytime contrast between land and sea can be impressive. Satellite picture 6D is an example of an early afternoon in spring with north winds and a high to the west of Ireland. The sea is largely clear of cumulus clouds whereas the land is well covered,

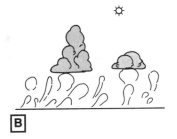

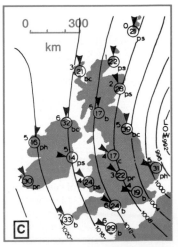

6B. Cumulus clouds form by blobs of warm air rising, often from sun-heated ground.

6C. A day with cold north-west to north winds and cumulus clouds. Widespread wintry showers of rain, snow and hail.

6D. Satellite picture of cumulus clouds on a spring afternoon in north winds, with marked contrast between land and sea. There is no relation between clouds and mountains. Some clouds are in rows along the wind, and there is drift out to sea. Showers of rain and hail are widespread over England.

particularly Ireland. See also 4B.

Some mornings start cloudless, and then cumulus clouds appear. Each cloud lasts only 5, 10 or 20 minutes, say, but the sky goes on looking much the same as new clouds take the place of old ones. The clouds evaporate by mixing with clear air around them, but sometimes tops will spread sideways into longer-lasting patches that may join to give a sheet over the sky - and so a promising day ends up dull (diagram 6E and colour photograph 6F). This is more likely to happen in a ridge or high rather than in a low.

The presence of sheets of cloud like stratocumulus or altocumulus may not stop sunshine setting off cumulus clouds, the more so if the sheets are shallow and it is summer. So cumulus and stratocumulus are commonly seen together (diagram 6G and colour photograph 6H).

When cumulus clouds first start to form, as small puffs, they sometimes do so in long rows more or less along the wind. Such *cloud streets* can also occur at any time during the day. Clouds in a street drifting across the coast over cool water will decay away. Satellite picture 6I shows rows of small cumulus clouds on a winter afternoon lying along the west winds over southern England and northern France. Clouds over England have tops about 2,000 - 2,500m, judged by their shadows. Another example of rows is satellite picture 11E over England west of the cold front.

6E. Stages in the spreading of cumulus cloud tops to form a stratocumulus layer.

6G. Mixtures of cumulus and stratocumulus are common.

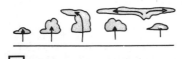

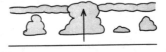

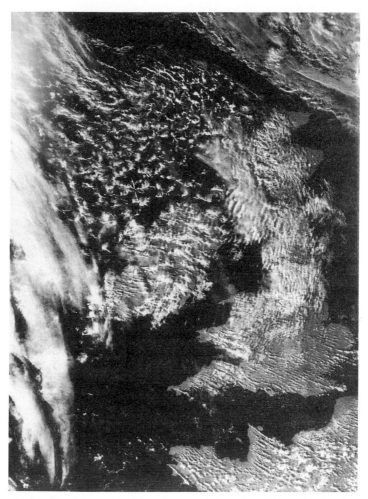

6I. Satellite picture of a winter afternoon. Rows of small cumulus clouds up to 150km long lie along the west wind over southern England and northern France. Tops up to 2,500m, judged by shadows. Because sunshine is weak, clouds do not form until the wind has spread some kilometres inland but, once formed, some rows stream off-shore. Anglesey seems to be not wide enough for clouds to form - unless they are smaller than the resolution of the satellite sensor. Note the complex pattern of spots and lines of cumulus clouds over the sea west of Scotland.

35

7. Cloudless Day

Days with little or no cloud are rare. In summer, they can give more than 15 hours of unbroken sunshine. Where is such weather to be found? Judging by Sections 5 and 6, it must be away from fronts and lows, and in airstreams too dry for much cumulus cloud to form. The most likely place is a ridge or high where the wind has come a long way over land. Map 7A shows a high to the east on the morning of a sunny summer day in the south-east that turned out hot. See also Map 4A and satellite picture 4B, which shows much of Britain basking in late summer sunshine. A long track from between south and east helps, but even north winds can be cloudless in winter over central England (map 7B), where in summer there would be much cumulus cloud.

When the pressure pattern changes so as to let in a moister wind, not only does the chance of stratus or stratocumulus cloud increase, but so does the chance of cumulus over land by day, the more so in the warmer months. Map 5E, for example, shows the kind of cloudy south-west winds that can spread across when a high that has been giving a

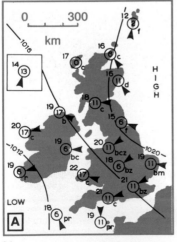

7A. With a high to the east and a low to the south-west, a warm and sunny morning across the middle of England before a hot day. Hazy east winds cover England and Wales. North-east coasts are cool with North Sea fog and drizzle. Rain and showers in the south-west as a low approaches slowly. (Inset shows Cairn Gorm summit.)

cloudless spell moves away. Sometimes the dry wind goes on blowing whilst a front edges its way in; the first clouds to end a spell of cloudless skies are then often patches of cirrus or altocumulus, followed later by dull weather near the front. Map 7C shows a high to the east with warm, largely cloudless south-east winds ahead of a cold front, and clouds creeping slowly east across Ireland and into western Scotland.

7B.   A winter high to the west gives cloudless north winds over central England. Some wintry showers in the north.

7C.   Largely warm, sunny but hazy south-east winds ahead of cloud, rain and drizzle on a slow, eastward-moving cold front.

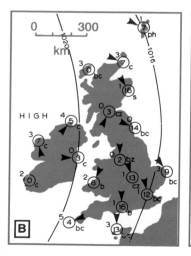

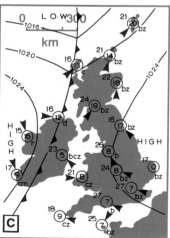

8. Hazy Day

Even a day with little or no cloud can be so hazy you are unable to see hills you know are only a few miles away. Those you can see are unusually blue or grey, and much of their detail is lost. The sky may be brownish white, and the sun glows like polished brass. Such view-spoiling weather can last for days.

Haze is almost always due to air pollution - smoke from chimneys, or smog from the effects of sunlight on vehicle exhausts. When winds blow gently over large urban or industrial areas, pollution gathers in the air, the more so where its upward spread is stopped at heights of, say, 1,000m. The main sources of air pollution over Britain are parts of central and northern England, and parts of central and eastern Europe. Hence west winds are often clean, whereas south-east winds can bring widespread haze, the more so when they are in a ridge or high. Satellite picture 8 shows haze streaming across Britain from continental Europe on the western side of a high centred over Germany.

If hazy air becomes moist, say by blowing across the sea or through a rain area, or it is cooled at night under cloudless skies, the haze can thicken so you can see no more than a few kilometres, the more so if there are low stratus clouds as well. The day is then very murky and the view is lost. Haze clears when the wind sets in from a new direction where there are fewer pollution sources. Behind a front there can be a quick change, so that familiar landmarks can be seen again. On very hazy days it is difficult to tell the kind of cloud in the sky, or its height. In map 7C, clean Atlantic air is edging slowly eastward to replace the haze that extends the length of Britain - from the Channel Islands to Shetland. Map 4A shows another example of haze reaching Britain with warm south-east winds; and map 14B shows warm, hazy south-east winds being replaced by clean westerlies behind a cold front.

Days with gales on the coast can bring another kind of haze - owing to countless salt particles formed by the evaporation of spray droplets from breaking waves. But salt haze seldom makes it impossible to see less than 10km.

8. Satellite picture about the time of map 7A. Widespread haze is streaming across France and the Low Countries to Britain. (To show this haze, processing the image has led to loss of details in the clouds.) Sea fog has spread over the north-east as far west as the Pennines, with tongues up Wharfedale (W) and the valleys of the Tyne (N) and Tweed (D). The North York Moors (M) stand above the fog top. Fair Isle (F) is making a wake in the fog.

Watch the weather map for persistent light south-east winds in a ridge or high bringing haze from continental Europe.

9. Foggy Day

Some mornings you wake to find yourself shrouded in fog. You may be able to see no further than about 50m, and trees are dripping wet. This fog may be all the more surprising because the night had been clear, calm and starry. The weather map shows a high, ridge or col with widely spaced isobars; the air is moist but there is little or no cloud. On such a night the ground cools quickly, and so does the air, until the moisture starts to condense - some as dew, and perhaps some as fog droplets. Drifting droplets collide with vegetation and may collect, only to run off as drips. By dawn the layer of fog may be 100m deep or more. As the sun rises in the sky, even though it may not be seen, the foggy air is warmed slowly and the droplets start to evaporate. By mid-morning in summer, but often later (if at all) in the weaker winter sun, the fog goes, sometimes through a spell of very low stratus clouds.

In clean air, fog can form in a few tens of minutes, and go away as quickly. In polluted air, the changes can take hours. Fog can also go if the wind picks up, or if a deep sheet of cloud spreads across to alter the heat balance of the fog.

Beware a clear, calm night. Be prepared for fog, but don't be surprised if there is only dew. Fog can thicken, or even appear first, just after dawn, as the sun starts to evaporate dew.

Windborne fog sometimes spreads in from the sea. Such fog can come by day or night when warm winds are blowing over a cool sea. It clears when the wind freshens enough, or changes direction so that drier air is brought in. Out at sea, fog can cover areas hundreds of

9. Sea fog spreading inland and dispersing over warm land.

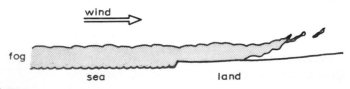

kilometres across (see satellite pictures 4B, 28B and 40B). Sea fog that has spread inland clears away much like night-time land fog (diagram 9; see also satellite picture 4B for clearance of sea fog downwind - west - of Prawle Point, Lizard and Scilly). Weather maps 4A and 7A show fog in Scilly and Shetland, respectively, with warm winds that have been cooled by the sea.

Dew can appear also during the daytime when moist air arrives after a cold spell, even when the sky is full of cloud. Roads and rocks then become unpleasantly wet.

10: Clear Day

Some days are gloriously clear. Far distant hills stand out against blue sky or sharp-edged clouds. The sky looks as though it has been washed - and indeed that's just what's happened. In the past few days the air has been through the vast cloud sheets of a low, where all the pollution particles had been trapped in the cloud droplets and fell out in rain. Then the air sank slowly in a ridge or high and reached us after blowing over no pollution sources, so it is almost as clean as it can be. Long tracks over the open ocean help to keep the air clean, so in Britain really clear days come with on-shore winds blowing from between south-west and north, often behind a cold front. On map 10A a cold front has just cleared the south-eastern corner of England, having crossed the whole of Britain from the north-west. Clean west or north-west winds have swept in from the Atlantic.

Clarity of the air is not cut down by clouds in the sky. Indeed, clouds pick out the clarity when they can be seen so easily much nearer the horizon than usual (diagram 10B).

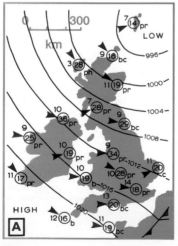

10A. Low to the north-east, and clean north-west winds have flooded across Britain behind a cold front that has just cleared south-east England.

10B. Great clarity of the air lets clouds be seen close to the horizon.

42

11. Rainy Day

Clouds are wet. Walking in cloud you gather water droplets on your hair and clothes. Plants also gather droplets as cloud blows through them.

Inside clouds, droplets knock into each other and grow slowly. It takes something like an hour to reach the size of a drizzle drop (which is still small enough to have no effect falling on a water surface), but not much longer to reach a rain drop. As a drop grows heavier it falls faster. Hence a rainy cloud not only has to last something like an hour, it must be deep enough to let drops grow as they fall. Drizzly clouds are usually at least several hundred metres deep; rainy clouds are much deeper. Most clouds are not deep or long-lasting so they cannot give rain.

Map 11A shows a low with a warm and a cold front. The warmest winds are the south-westerlies between the two fronts - in what is called the *warm sector* of the low. In a warm sector there is often much stratus and stratocumulus cloud with outbreaks of drizzle (like colour

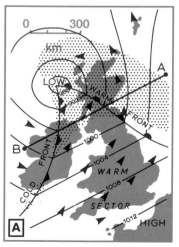

11A. A 'warm sector' low, with areas of rain shown stippled. Compare with maps 5B and 16F.

11B. Vertical cross-section through a warm sector, along the line AB of map 11A, showing typical layering of clouds and areas of rain.

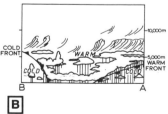

photograph 5A). Near the low centre there are often hidden upper clouds with outbreaks of rain. Ahead of the warm front, there is a belt of rain 200km wide or more, moving across country with the front. Near the cold front there is usually a narrower and heavier band of rain. Fronts vary greatly: some have little or no rain. If a front is coming towards you, check with the forecast how heavy and how long-lasting the rain is likely to be. Map 5B shows a warm sector, with the low centre moving east towards northern Scotland.

We can see there is a sequence of weather as a warm sector low and its fronts pass overhead (diagram 11B). In a ridge ahead of the low there are likely to be cumulus and stratocumulus clouds by day (like colour photograph 6F), largely clearing at night. As the wind backs to south before the warm front, upper clouds spread from the west. The first to come are likely to be cirrus and cirrostratus, perhaps with a halo - a white ring around the sun or moon at 22°. Then come altostratus and altocumulus (like colour photographs 5G and 5H), darkening as they deepen to *nimbostratus*, from which rain falls for several hours (colour photograph 11C), with lowering stratus clouds beneath. After the wind veers as the warm front passes overhead, there are the warm, damp south-west winds and the clouds of the warm sector. As the cold front comes near there can be outbreaks of rain from unseen upper clouds. Rain can stop before or after the wind veers again at the cold front but, as drier air spreads in, the cloud base rises, and breaks grow as the cloud changes to cumulus in the west or north-west winds. The whole sequence may take one to three days, and the details can vary a lot from one warm sector to another. For examples of rain at warm fronts see maps 3D, 5B, 15I, 31B and 31D; for rain at cold fronts see maps 3A, 3E, 5E, 7C, 12F and 15E.

Signs of likely rain to come are: sheets or cirrus, cirrostratus or altocumulus spreading across the sky, thickening and darkening within a few hours to altostratus. Watch out, too, if formerly rainless sheets of stratocumulus darken and lower. Always remember there may be unseen upper clouds that can give rain falling through the lower clouds.

The combination of rain and wind is irksome and can be dangerous if wind-chill is extreme. Such driving rain comes most often with winds blowing from between south-east and west, with the warm and cold fronts of lows typically moving between Scotland and Iceland. It can be almost continuous for days or even weeks in western and northern Britain as one low after another passes by. Watch for large, deep lows forecast to come near or over the British Isles. Four satellite pictures illustrate some typical events. 11D shows a deep low approaching from the west; later it moved north-east across Ireland and Scotland, giving 4in of rain in Snowdonia and ending a long spell of summery weather. 11E shows a deep low centred between Orkney and Shetland giving rain everywhere and winds over northern Scotland reaching force 11 at sea level, with a gust of 153mph recorded on Cairn Gorm summit. 11F shows another deep low, this time over southern England and giving a wet and stormy day over England and Wales. 11G shows an old low off southern Ireland that crept eastwards to the North Sea over the next three days, giving periods of rain almost everywhere.

11D. Deep low to the west of Britain - to help orientation, Orkney (O) and
Cornwall (C) are highlighted. On the next day the low centre moved
across Scotland, bringing over 4in of rain to Snowdonia. Note the
thickening of aircraft contrails near the north-western edge of the
frontal cloud.

11E. Deep low centred between Orkney and Shetland. Extensive rain clouds have moved away to the north and have already been followed by showers into Ireland, Wales and northern England (as well as Scotland, where they are partly hidden by streaks of cirrus clouds). Over Ireland, the shower clouds have fuzzy tops leaning northwards.

11F. Swirl of frontal clouds around a deep low centred over southern England. Wet and stormy weather over England and Wales. Scottish snow showers later spread south. Some land is visible in the cloud gaps: parts of Scotland (Lewis and Kintyre) and Ireland (Galway Bay and Waterford-Rosslare).

11G. Old, slow-moving low off south-west Ireland. It took three days to
reach the North Sea, giving rain on the way almost everywhere.
The spiral band of frontal cloud has broken into a complex pattern
of cumulus and upper clouds. (Try describing this, let alone
forecasting its change!). Some land is visible: for example, Mull (M),
Jura and Islay (J), Kintyre, Merseyside, Pembrokeshire, Somerset,
Devon and Kent (K).

12. Showery Day

Rain sometimes comes in short-lived bursts lasting much less than an hour, with bright or sunny spells between that help us to dry out. Unlike the steadier and more lasting spells of rain that fall from widespread sheets of cloud near fronts and lows, these *showers* come from much smaller clouds called *cumulonimbus*. You can tell these clouds by their tall, swelling, domed and fibrous tops, shining brightly in the sun (diagram 12A). As a cumulonimbus cloud starts to rain out, its hard-edged cauliflower-like top become streaky and perhaps stretched across the sky in a plume or anvil shape (diagram 12B and colour photograph 12C). But when skies are more cloudy the tops may be hidden and we must look for the curtains of rain falling from them (diagram 12D and colour photograph 12E).

Shower clouds grow from cumulus clouds. Over land they are commonest in the warm months and at the warmest time of day. If you see cumulus clouds building up quickly during the morning, watch out for showers by midday. They are likely to be heaviest in late afternoon, but die away in the evening. Over sea and windward coasts they are most common in airstreams coming quickly from high latitudes around a large and deep low. Map 12F is an example with showers of rain, hail and sleet in strong west winds south of a deep low to the north of Britain. The corresponding satellite picture (12G) shows that the west winds north and west of Scotland are full of shower clouds. (For examples of showery days, see maps 3A, 3B, 3E, 6C, 15A and 15E.) Such showers fall by day and night. Nearer highs, cumulus clouds are usually too shallow to grow into cumulonimbus and they give no more than light showers (see map 7B).

Sometimes showers are not widespread, as they are over Scotland and northern Ireland in map 12F, but concentrated in bands with few elsewhere. Two satellite pictures illustrate this: 12H shows two long bands over southern Britain, forming downwind of peninsulas in west-south-west winds; 12I shows a single band forming near the North Channel in north-north-west winds and stretching to mid-

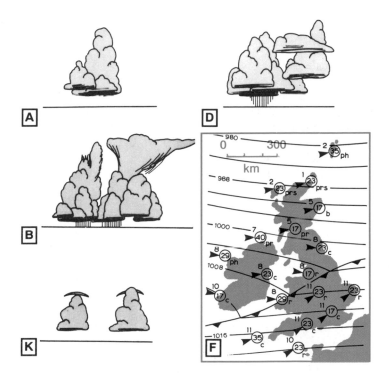

12A. Massive cumulus clouds, beginning to change into a *cumulonimbus*.

12B. Cumulonimbus clouds with tops becoming streaky (left) and anvil-shaped (right).

12D. Growing cumulonimbus hidden by cumulus in the foreground, but shown up by its distant dark base and curtain of rain.

12F. A cold front with rain is moving south-east across England. A deep low to the north gives strong west winds and wintry showers of rain, sleet and hail over Scotland and Ireland, but there is clear sky to the east of the Highlands.

12K. Cloud cap, through which the swelling cumulus top soon bursts - a sign of likely showers to come.

12G. Satellite picture for about the time of map 12F. To the north and west of Scotland there are hundreds of large wintry showers in an array of blobs and rings, with fuzzy tops leaning to the north-east, but downwind of the Highlands there is none.

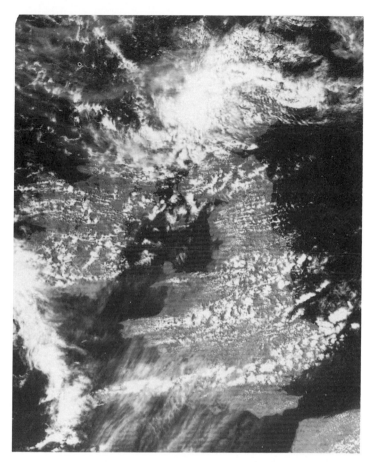

12H. Sometimes showers form in rows along the wind - similar to 6I but on a larger scale. On this spring afternoon there were two main bands: one from Cornwall to north of London (partly hidden by cirrus clouds), the other from east Wales to the north Midlands. These bands had formed downwind of peninsulas; other showers have been taken over the North Sea.

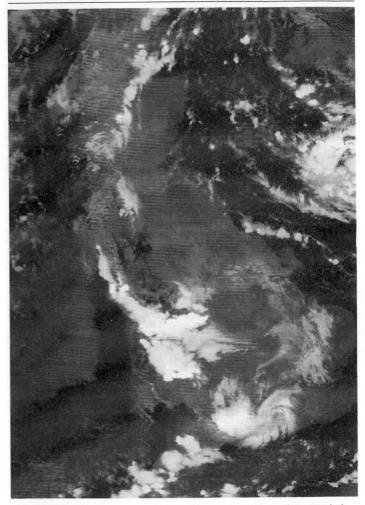

12I. A winter band of heavy showers forming in north-north-west winds near the North Channel and stretching to mid-Wales, causing floods, thunder and snow there.

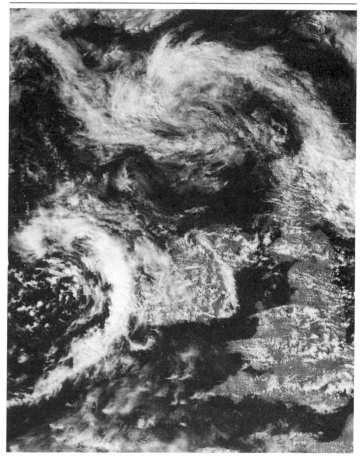

12J. Early afternoon in summer. An old low centred north-west of
Scotland drifts slowly eastwards. In contrast to the scattering of
showers over land (see also 11E, 11F and 12G), there is a large
comma-shaped cluster west of Ireland moving east to give heavy
rain and thunder over southern Britain during the next 24 hours.

Wales, where there was local flooding. Over the ocean, too, showers can become concentrated, sometimes into a large comma-shaped mass that can give a spell of heavy and more prolonged rain, along with a trough of low pressure (12J is an example spreading into the south-west).

Always keep an eye on cumulus clouds upwind. Watch for:

- rain curtains cutting the visibility down to a few kilometres;
- rainbows (if the sun is downwind);
- large cumulus, forming dense, streaky, spreading tops;
- cloud caps through which the swelling tops soon burst (diagram 12K);
- sudden, squally winds blowing outwards from dark-based clouds.

After several wettings from passing showers it is helpful to know when they are likely to stop. Look for a change of cloud type: from cumulonimbus to shallower cumulus and stratocumulus, say towards sunset or as a ridge of high pressure moves over you.

13. Dry Day

Many days give no rain. The clouds are then either too shallow (say, sheets or patches of stratus, stratocumulus or altocumulus - see colour photographs 5H, 6F, 6H and 37F) or too short-lived (say, small cumulus - see colour photograph 6A). If morning cumulus clouds stay flat, or their tops build up to hardly more than twice their base height, showers are unlikely. If cirrus and cirrostratus clouds spread across the sky but only slowly, the more so if they are not followed by altocumulus or altostratus, rain is unlikely perhaps until the next day at the earliest. If stratocumulus clouds stay broken, or they stay above 1,000m even if there is a complete cover, there may be no more than patchy drizzle.

Highs and ridges often give dry days (see, for example, maps 3C, 4A, 7A and 7C); even fronts within them can give little or no rain because their clouds are too shallow, although in winter light rain or drizzle is possible from clouds with tops no higher than 2,000m. After a dry spell of a few days, watch the weather maps for fronts coming along (maps 7C and 14B). At all times watch for winds changing direction to blow from low latitudes across the Atlantic - they can bring grey, drizzly weather and so end a dry spell.

Sometimes highs are very slow-moving near Britain for many days, even several weeks. These are called blocking highs, because they block the movement of lows, which then travel further north (or south) than usual. When this happens there is a long spell of dry weather, often warm or hot in summer, but foggy or dull in winter.

14. Thundery Day

Thunder is rare in Britain: most places have only 5 to 10 days a year, on average. At worst it is only noisy, but lightning can be a danger. Thunder is the noise made by giant electric sparks we call lightning. Because sound travels much slower than light, the time between flash and clap is greater the further away is the flash - 3 seconds for every kilometre. So, a 12-second time gap means the flash is 4km away, and if the thunder cloud is coming towards you on a 30kph wind it will be overhead in 4/30 of an hour (8 minutes). Thunder can seldom be heard

A

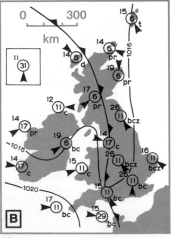

B

14A. Massive group of showers, likely to give thunder and hail.

14B. Summer cold front moving slowly east with severe thunderstorms. Torrential rain, large hail and sudden temperature falls ended a spell of hot, hazy weather, and three were killed by lightning. (Inset shows Cairn Gorm summit.)

more than 20km from a storm, but lightning may be seen more than 100km at night.

Lightning tends to strike upstanding things like trees and tall rocks. If they are wet with rain, the electric current is likely to pass through the water; but if they are dry on the outside, lightning can pass through moisture in the rock cracks, or through tree sap, sometimes explosively because the moisture boils at once, splitting open the rock or tree.

Thunderstorms usually grow out of large and energetic showers (diagram 14A). They often come with hail. If you see cumulonimbus clouds growing large, the more so if they are giving heavy rain or hail, watch out for lighting and thunder. A cloud often gives thunder for less than half an hour, but as new showers grow and pass by on the wind a storm can last for hours at any one place.

Over land, thunder is most likely in the afternoon, on days with plenty of showers (Section 12). More widespread thunder, with outbreaks of heavy rain, can come by day or night in summer when a cold front from the west has become slow-moving after a warm spell (map 14B), or when a low or trough is slow-moving over or near Britain (map 14C). The satellite picture 14D, about the time of map

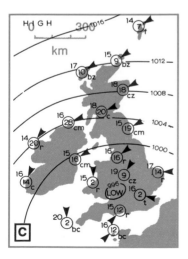

14C. Slow-moving summer low over southern England with widespread thundery rain. Mostly cloudy east and north-east winds over Scotland and Ireland, but sunny in the far north with some sea fog.

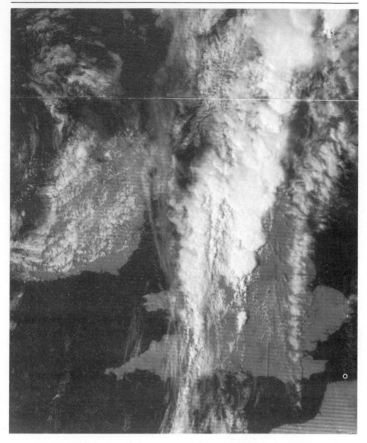

14D. Satellite picture about the time of map 14B. Large thunderstorms
     moving from the south are at various stages of formation, and their
     anvil tops cast shadows to the north-east. Over southern Scotland,
     these tops tower above the cirrus clouds.

E

14E. *Castellanus* cloud - a sign of likely thunder to come. Most of the clouds over south-east England in 14D are castellanus and dense cirrus.

14B, shows a north-south line of severe afternoon thunderstorms along a cold front.

Some winter showers give a few claps of thunder, the more so with west winds in western and northern Britain.

Watch out for a cloud called *castellanus*; it looks like rows of turrets sprouting from a common base (diagram 14E). If these clouds spread and grow along with others as the front, low or trough gets near, there is a fair chance of thunder to come.

15. Snowy Day

Falling snow makes a pretty scene, but life outdoors becomes rough, progress is slowed as dangerous hollows get blanketed, and things may be hard to see only a kilometre or two away.

The kinds of weather patterns that lead to snow are much the same as for rain; it's simply that the air is cold enough to let snow reach the ground before it can melt. Long spells of snow come with lows and fronts, and snow showers come with cold airstreams. Winds can blow from any direction, but those from between south and west seldom bring snow for more than a few hours.

Winds from between west and north, with a large and deep low to the north or over the North Sea, can bring heavy snow showers to west- and north-facing coasts and hills (such as in maps 3A, 3B, 3E, 6C and 12F), but cold east winds can give snow showers even with high pressure (often centred over Scandinavia). Map 15A shows snow showers spreading to Scotland on north winds around a high just south of Iceland, whilst a cold front over England and Wales is bringing rain that freezes on the ground as *glazed frost*, and up to 6in of snow

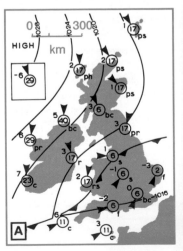

15A. Showers of snow and hail have reached Scotland in north winds behind a cold front over England and Wales that is creeping south against persistent freezing fog. (Inset shows Cairn Gorm summit.)

15B. Satellite picture about the time of map 15A. Innumerable small snow showers are spreading into Scotland whilst a cold front lies across England and Wales, where the highest (whitest) clouds occur.

CAPTIONS FOR COLOUR PHOTOGRAPHS

5A A sheet of almost uniform *stratus* cloud, base about 200m. Risk of drizzle at any time. Cornwall.

5G Multilayered upper clouds ahead of a front approaching Connemara. *Cirrostratus* thickening to *altostratus* through which the sun shines dimly. Risk of steady rain within a few hours if cloud goes on thickening and darkening.

5H Shallow, broken sheet of banded *altocumulus* cloud at about 4,000m over Connemara. Thickening and lowering of such cloud is often associated with approaching fronts and lows.

6A Many small, shallow *cumulus* clouds, showing typical flat bases (here at about 1,200m) and rounded tops. Above them are thin, high, fibrous *cirrus* clouds, thickening from the west (left) into milky *cirrostratus*. Typical of a ridge of high pressure ahead of a low coming from the west. Risk of steady rain within a day if high clouds thicken and lower. Macgillicuddy's Reeks, Kerry.

6F *Cumulus* clouds, based at about 1,000m and just touching Snowdon. Tops are beginning to spread at about 2,000m to form shelves of *stratocumulus* beneath a temperature inversion. Risk of day becoming dull if shelves gather into an unbroken sheet.

6H *Cumulus* towers (base about 1,000m and a result of sunshine heating the land) beneath a broken sheet of *stratocumulus* cloud (at about 2,000m and a result of gentle lifting ahead of an approaching low). Risk of light showers if cumulus clouds grow larger; risk of drizzle if stratocumulus clouds thicken and lower in the next few hours. Donegal.

11C A mixture of clouds: patches of *stratus* (base about 200m) and *stratocumulus* (about 500m); and almost uniform *nimbostratus* well above the tops of the Cuillin. The lake surface shows it is raining but there is little or no wind. A cold front has passed not long since, and low clouds have broken rapidly. Upper clouds are still deep enough to give rain but the lighter patches show that the expected clearance is already under way. But watch for cumulus clouds and showers within a few hours.

12C *Cumulonimbus*, with dark base at about 1,000m and anvil-shaped, streaky top at about 10,000m, accompanied by many smaller cumulus clouds as well as patches of altocumulus. The large cumulus tower over-topping the anvil shows strong growth that will keep the shower going. This shower is moving away, so watch for more coming from behind. Pembrokeshire.

Above: 5A
Below: 5G

Above: 5H
Below: 6A

Above: 6F
Below: 6H

Above : 11C
Below: 12C

Above : 12E
Below: 21A

Above: 21D
Below: 37C

Above: 37F
Below: 37I

Above: 38C
Below: 42B

CAPTIONS FOR COLOUR PHOTOGRAPHS

12E   A showery day. A mix of large *cumulus* clouds (dark bases, around 700m) and *cumulonimbus* (tops not seen, but shown by the distant curtain of rain hiding mountains around the head of Loch Linnhe). Keep looking upwind for signs of more showers coming.

21A   *Stratus* cloud forming in shreds at about 400m as the wind blows up the crags from the left. The shreds are forever forming in the same place and merging to give hill fog. Watch the cloud base: if it lowers steadily, expect widespread stratus cloud and drizzle; if it rises, expect the tops to clear and even breaks in the sheet. Snowdonia.

21D   *Stratocumulus* cap, base about 500m, as the wind blows from the left over the Rivals. Above, an almost uniform sheet of *altostratus* cloud has a base about 2,000-3,000m. Watch for darkening and lowering of the upper cloud - it may lead to steady rain. Watch the base of the lowest cloud - it may herald moister air, low stratus cloud and drizzle. Snowdonia.

37C   A cold front has passed through. *Cumulus* clouds, with base about 900m, are coming from the right, but patches of lower, pre-frontal clouds linger in the slowly-overturning eddies sheltered below the crag tops. The patches should go in the next hour or two as they mix with drier air blowing over the ridge. Snowdonia.

37F   Lens-shaped and banded *wave clouds* over Kerry, base about 1,500m. Small cumulus clouds beneath, base about 800m. Watch the wave clouds: if they change little, the day will stay dry; if they spread and thicken into a lowering sheet, there may be drizzle.

37I   Fitful sunshine on a day with two, thin, broken sheets of *stratocumulus* cloud, one above the other. The wind blows down the slope from cloud over the Glyders, on the right, through a *wave gap*, and rises into the dark *wave cloud* on the left. Snowdonia.

38C   *Stratus* cloud pouring through the narrow head of the Llanberis Pass. The flat top marks the base of a temperature inversion, beneath which the cool, cloudy air flows down-slope and warms by compression. There can be a big contrast between nearby valleys when a cloud sheet is blocked. Watch those patches of cirrus cloud above to see if many more spread across the sky, perhaps ahead of a front or low. Snowdonia.

42B   Thick, lumpy *stratocumulus* clouds with ragged base about 400-600m and curtains of drizzly rain blowing off Brandon Mountain, right. If the base rises steadily, expect the drizzle to die away. Kerry.

as it battles southwards against freezing fog that had persisted for several days - what a mixture! Satellite picture 15B shows the countless small showers spreading south into Scotland, whilst the cold front lies across England and Wales.

Map 15C shows cold air coming from both east and north. With a low over France, east winds cover southern England bringing light snow showers to the Midlands. At the same time, with a high west of Iceland, north winds and snow showers are about to reach northern Scotland behind a cold front bringing more general snow (see satellite picture 15D). Even south-west winds can bring snow showers if the air is very cold before it leaves North America to cross the Atlantic. Map 15E shows the *Braer* storm of 1993, centred to the north-west of Scotland. It was the deepest low on record for the North Atlantic, when exceptionally cold south-west winds brought heavy snow showers over Scotland, disrupting transport and power lines. A gust of 120mph was recorded on the island of North Rona. Satellite picture 15F shows the impressive swirl of frontal clouds around the low causing the *Braer* storm. Snow showers are streaming eastwards across northern Scotland; and the tufted northern edge of the huge cloud mass is cirrus - often found ahead of a front (see colour photograph 5H; also satellite

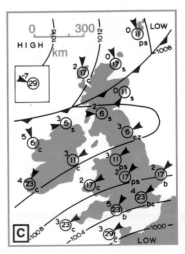

15C. Cold air reaching Britain from both north and east. Snow showers are about to reach northern Scotland behind a cold front bringing wide-spread snow to southern Scotland and northern Ireland that caused considerable disruption on roads and rail. Strong, cold east winds over England bring a few snow showers. (Inset shows Cairn Gorm summit.)

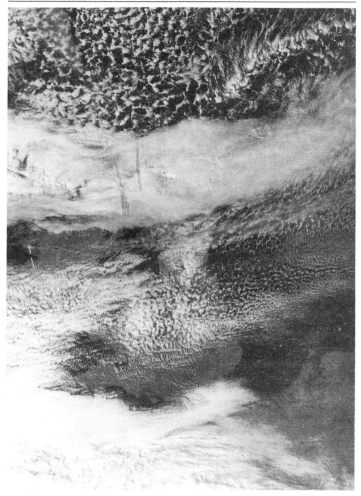

15D. Satellite picture about the time of map 15C. There are small snow
showers over the Midlands, and further snow showers are about to
reach northern Scotland behind the cold front cloud band.

pictures 16I and 31E).

Showers at night in very cold east winds form only over the sea but they can bring heavy snowfalls to windward coasts. Satellite picture 15G shows along-wind rows of shower clouds forming in sub-zero air once it has spread some tens of kilometres out to sea to the west of Ireland, Wales, south-west England, northern France and the Low Countries. Note how the clouds get larger downwind. The lie of such bands changes with wind direction, and if they swing just a little so as to cross the coast then snowfalls there can be heavy. Later on this day there was over a foot of snow in east Kent, and the next day had widespread drifting over southern England.

A low moving east over England can bring long-lasting snow and east winds to Scotland, but if the low crosses northern France then England and Wales have the falls. Map 15H is an example.

Often only a small change in temperature will mean a difference between snow and rain. With south-east winds ahead of a slow-moving front, snow can go on falling for more than a day (map 15I), slowly changing to sleet and then rain as the front comes close.

If you look at a snow flake you will see that it is a loose jumble of fragile ice crystals, many like six-pointed stars. These crystals grow

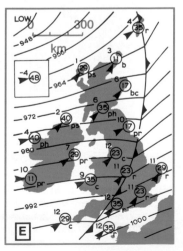

15E. The storm responsible for the final break-up of the oil tanker *Braer* in 1993. This low was the deepest on record for the North Atlantic, bringing unusually cold south-west winds and wintry showers of snow and hail to Ireland and Scotland behind the fast-moving cold front with its rain. (Inset shows Cairn Gorm summit.)

15F. Satellite picture about the time of map 15E. The swirl of frontal cloud is impressive, and tallies with the severity of the winds. Snow showers are rushing across northern Scotland, with their anvil tops leaning northwards. Caithness and Orkney (O) are just visible in the clear slot west of the frontal cloud band, where sinking air curls into the low centre. Cornwall (C) and Kent (K) lie under the frontal clouds.

15G. Bands of snow showers lying along very cold east winds. Windward coasts can get heavy falls. Showers along the east coast of England are partly hidden by cirrus clouds. Showers from the Irish Sea decay into cirrus patches (old anvil tops) as they drift across Ireland.

inside the clouds and tangle together as they fall. In very cold weather the flakes are small, but near melting point they are sticky and sometimes join into large clumps.

Snow can fall when the air is warmer than melting point. This happens when the air is dry, for then the flakes slowly evaporate as they fall and the resulting cooling keeps them below melting point. Snow showers in dry north winds can be seen at temperatures as high as 7°C.

Beware: if rain is forecast but the air is cold and dry, snow may fall instead.

15H. A winter low centred over northern France giving widespread snow over England and Wales.

15I. Slow-moving winter warm front bringing long-lasting snow, turning to sleet, rain and drizzle near the front. Cold south-east winds being replaced by mild south-westerlies.

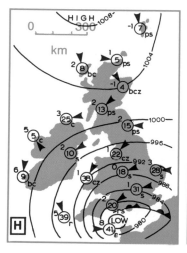

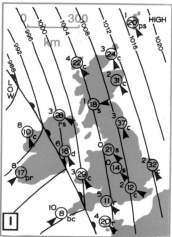

16. Blizzard

Snow and strong winds can lead to a bad day. The wind blows the already fallen snow into swirling clouds - a *blizzard*. Going forward becomes very hard in the choking, blinding spindrift. When freshly fallen, snow is loose and easily blown in a cold, dry wind little stronger than force 3. Older snow, more granular and perhaps crusted after lying a few days, drifts less readily. Because the wind is calmer in the lee of obstructions like plant tufts and rocks, blowing snow tends to settle there and form drifts stretching downwind (diagram 16A). In time, these drifts grow until the wind blows smoothly past them. Just in the lee of a ridge, swirling eddies keep the ground more or less free of lying snow for a while, but even there a long-lasting blizzard can lead to a smooth drift that may be metres deep. Although such a place might be chosen as a shelter from the wind, there is the risk of being buried later. Alongside upstanding rocks, strengthened winds keep the ground largely free of snow (diagram 16B), whereas even in front of them there can be places where the wind weakens enough to let a drift gather (diagram 16C). Photograph 16D shows scouring around boulders and a lee-side drift, and 16E shows countless small saturated

16A. Stages in the growth of a snow drift behind a boulder.

16B. Plan view of a boulder with windswept scoops on either side.

16C. Drift forming upwind of a boulder.

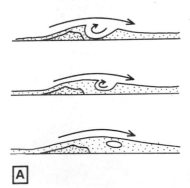

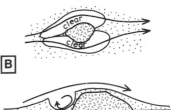

16D. Wind scours around a boulder.

16E. Countless small drifts in the lee of tufts of grass and heather.

drifts in the lee of tufts of grass and heather.

Be on the watch for a blizzard when both snow and strong winds are likely. Map 16F shows an almost stationary warm front lying across the English Channel, with sub-zero east winds all day on its north side (the temperature had fallen to -27°C in Scotland the previous night). Frontal snow was widespread over southern Britain for up to two days, the more so in south Wales, which had 1m (3ft) depth widely and was almost cut off for three days by drifts reaching 6m (20ft), but Scotland was clear (satellite picture 16G). On the next day, the warm west winds rising over the front gave rain in south-west England that froze as it fell into the cold easterlies beneath.

A bad day can be expected when snow falls on the cold side of a deep winter low moving slowly east across Britain. Map 16H shows a deep low crossing southern Scotland - the infamous Burns' Day storm of 1990. Rain and showers are falling almost everywhere. Moreover, it is not surprising that the rain and sleet with such low temperatures turned to snow as the low moved away eastwards. It gave a blizzard in Scotland with up to a foot of level snow. Severe gales over England and Wales, with gusts over 100mph, caused widespread damage and 47 people were killed. Satellite picture 16I shows this storm, with a vast

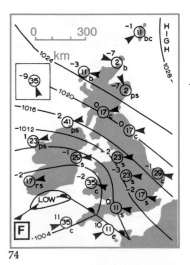

16F. An almost stationary warm front across the English Channel, separating sub-zero east winds from mild westerlies. A blizzard was widespread for up to two days over southern Britain, with drifts in south Wales up to 6m (20ft). Note the strong winds and lack of snow to the west of the Welsh mountains. (Inset shows Cairn Gorm summit.)

16G. Satellite picture about the time of map 16F. Frontal clouds in the south are giving much snow, but north-west Scotland is largely cloud-free.

swirl of frontal cloud around the low centred over southern Scotland. The whole of Britain is cloud covered.

Even one heavy shower in a strong, cold, dry airstream can turn a fine and bracing day into a blizzard (for example, map 15E).

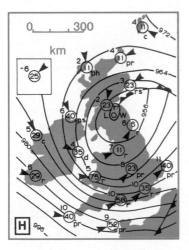

16H. The Burns' Day storm of 1990. A deep low is moving eastwards across southern Scotland bringing cold rain and sleet that changed to a blizzard there. The widespread severe gales over England and Wales go with the very close isobars. (Inset shows Cairn Gorm summit.)

16I. Satellite picture about the time of map 16H. The vast swirl of frontal cloud over Scotland is giving rain and sleet, whilst showers over England and Wales are hurtling eastwards in severe west winds causing widespread damage and loss of life. L marks the low centre, and the positions of Shetland (S) and Kent (K) are also shown. (This is a mid-winter picture so the sun is very low in the north and shining on the sides of the clouds, not the tops.)

PART THREE
Weather on the Tops and in the Valleys

In the remaining Sections we look at those days when weather in the mountains differs from that over low country. This often happens, so when watching, listening to, or reading a forecast for the general public it is necessary to make some allowance for the differences. Sections 17 to 31 give some guidance on how this can be done for mountain tops, and Sections 32 to 46 for valleys. Automatic telephone forecasts for mountainous areas draw attention to some of these differences.

17. Windy Top

Mountain tops are almost always windier than open, low country. There are two reasons for this. Firstly, even over open country the wind very often strengthens with height. Secondly, the wind strengthens as it blows around and across mountains.

The strengthening with height can be seen easily in the shapes of shallow cumulus clouds forming in the morning: they often tumble forwards as the wind pushes the upper parts of masses of rising warm air faster than the lower parts (diagram 17A). Narrow cumulus clouds can become leaning towers diagram 17B). Because there is this strengthening with height, the wind speed on an isolated 1,000m *peak*, where the wind tends to blow around rather than over, can be twice the speed over open, low country. The speed might be judged from a valley by the motion of cloud patches or their shadows, or even the blowing of spindrift.

The wind blows over rather than around a mountain *ridge*. On many days it does this in a way like water flowing over a boulder in a river. Just as the water speeds up as it becomes shallower over the boulder, so the wind strengthens, the more so when the lowest clouds are stratus or stratocumulus, not cumulus (diagram 17C). On such days the wind over a ridge can be three times the speed over open, low country, and it may be particularly gusty.

To illustrate the windiness of the tops, speeds at the Cairn Gorm summit automatic weather station (1,245m) have been added to some of the weather maps in this book. Speeds are often two or three times those over open country, even with on-shore winds.

Look for these strongest ridge-top winds on days with much low cloud, as in a warm sector of a low (maps 5B and 11A); or when a warm front is coming towards you and is less than about 100km away (maps 3D and 16F); or just behind a cold front, where the cold air has deepened to rather more than the height of the ridge; or around a strong high. The strongest winds in a warm sector tend to be in a long belt about 100km wide just ahead of the cold front (map 27A). When the winds in this belt are forecast to be strong or gale force over low ground, they can be hurricane force and very dangerous over high ridges. On showery days, however, with large cumulus and cumulonimbus clouds, the wind blows more easily over ridges. The strengthening with height is then less but still needs to be heeded, the more so when winds are already strong over low country (as in maps 3A, 3B, 6C and 12F). In map 15E, although the gales of northern Ireland have not yet reached western Scotland, Cairn Gorm summit already has near-50mph winds.

Among pinnacles, the wind can be very erratic as gusts make their way between them and as short-lived eddies form and are shed downwind. Winds there resemble those blowing between tower blocks in a city.

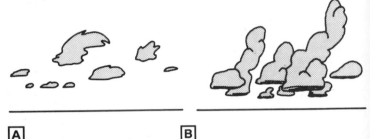

A   **B**

17A. Forward-tumbling small cumulus clouds when the wind strengthens with height.

17B. Leaning cumulus towers.

17C. Strengthening of winds over mountains because lifting is less at greater heights.

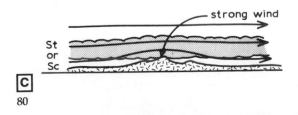

C

18. Calm Top

Mountain tops are not always windier than open, low country. There are two reasons for this. Firstly, the wind can sometimes weaken with height even over open country. Secondly, an eddy with weak winds can form in the lee of a ridge.

A weakening of wind with height is most likely above about 500m, but it happens on only a few days, for example in east winds on the south side of a high moving eastwards.

There is often a leeside eddy when the wind blows across a crested ridge (diagram 18A). The wind blows up the windward slope but leaves the ground at the crest. On the leeward slope the wind blows in the opposite direction, often weakly and fitfully, and the junction between the two winds can be very sharp. Much the same thing can happen at the windward edge of a flat-topped ridge (diagram 18B), or even on a windward slope where the steepness changes suddenly (diagram 18C). With a rounded ridge, the eddy starts a little way down the leeward slope.

These quieter eddies in the lee of knolls provide convenient resting places, and when snow is blowing along the ground they can be sought out for making a shelter, at the same time remembering that deep drifts can build up there. Indeed, on the leeward side of a ridge the drift becomes an overhanging cornice (diagram 18D).

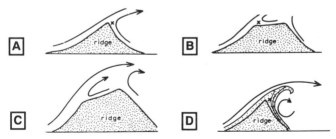

18A. Lee-side eddy, with light winds at X, just in the lee of the ridge crest.
18B. Eddy at upwind edge of flat-topped ridge.
18C. Eddy in the lee of a change in slope.
18D. Snow cornice overhanging leeward side of a ridge.

19. Cold Top

Even though a forecast is for warm weather, it must not be forgotten that it nearly always becomes colder the higher we go. The lingering of high-level snow is a reminder of this cooling with height. The rate of fall of air temperature with height is known as the *lapse rate* - on average it is about 1°C for every 200m. Using this lapse rate and the valley bottom temperature we can estimate the temperature at any height above the valley. For example, suppose the temperature in a valley at 300m above sea level is 10°C; what is the temperature on the summit at 900m? The height difference is 600m, so the temperature difference is 3°C; and the summit temperature is therefore 10 - 3, or 7°C (diagram 19A).

To illustrate the coldness of the tops, temperatures at the Cairn Gorm summit automatic weather station have been added to some of the weather maps in this book. Note how low the temperatures are when it is already near freezing at sea level (maps 3E, 15A, 15E and 16F). In the last example, with -9°C and 35mph, the wind-chill is so severe that exposed flesh would freeze. Map 15E is almost as severe. Even during a hot afternoon over low ground, wind-chill can make the tops feel almost cold if the wind is strong enough - for example, map 14B.

Inside clouds, the lapse rate is seldom more than about 1°C for every 200m, but beneath clouds it can be as large as 1°C for every 100m. Windy days and cumulus days are likely to have such a large lapse rate, for the air is then well stirred. Take an example on such a day. Suppose the temperature is 7°C in a valley at 200m above sea level on a sunny but windy winter afternoon. At 1,200m the temperature would be about 7 minus 10, or -3°C (diagram 19B). Thus, the summit would be below freezing and, bearing in mind the likely strengthening of wind with height, the weather there would be severe, despite the pleasantness of the valley, the more so if showers are blowing along in the wind.

On this day, 0°C would be reached at about 900m. This height is known as the *freezing level*; it can vary in height greatly from day to day, and is on average lower in winter than in summer, and in northern Britain rather than the south. The height of the freezing level is included in the Mountaincall recorded forecasts. Winter temperatures are lowest in strong easterly winds, and some days may be no warmer than -10°C at 1,200m.

When the wind blows across a ridge, summit temperatures can be lower than in air at the same height above low country upwind. This is most likely where a sheet of stratus or stratocumulus cloud has been deepened over the ridge. Although the cooling is usually small it can lead to unexpectedly icy weather when temperatures would otherwise be just above freezing.

The difference in air temperature between day and night on the tops is less than in the valleys. Indeed, in persistent hill fog, the temperature may change little over many hours.

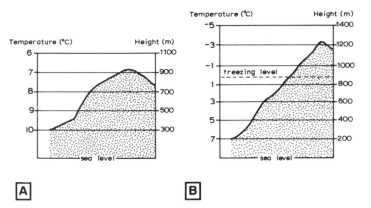

19A. *Average* temperature lapse rate: 1°C fall for every 200m gain in height.

19B. *Large* temperature lapse rate: 1°C for every 100m gain in height.

20. Warm Top

Sometimes a mountain top is surprisingly warm judged by the forecast for low ground. There are three main ways in which this can happen.

1. After a windless night with little cloud, instead of the air being cooler at greater heights it is warmer. We say there is a temperature *inversion*. Quite often the temperature can *increase* by 5°C in going up only 100m, and a more usual lapse rate may not start until a height of 200m or 300m above the valley (diagram 20A). The effects of a *ground inversion* can sometimes be seen on cold mornings when white hoar frost lies only on the valley bottom and not the higher slopes. Estimating summit temperatures is not easy when there is a ground inversion because you will not know how strong or how deep it is. Suppose there is an inversion of 5°C over a depth of 300m, and we use the first example in Section 19, ie. 10°C in a valley at 300m above sea level; what is the temperature at 900m? Given this inversion, the temperature at 600m is 10 + 5 = 15°C. Then taking a lapse rate of 1°C for every 200m rise, the temperature at 900m is 15 - 3 = 12°C. Contrast this with 7°C for the example in Section 19.

2. On some days, even windy or cloudy ones, there can be an *inversion aloft*. Diagram 20B shows an example where there is an inversion of 3°C between 600m and 1,000m. Both above and below this inversion layer the lapse rate is about average. Unlike a ground inversion, which usually forms on a clear, calm night and goes away as the morning gets warmer, an inversion aloft can occur by day or night, and it is not easy to tell if one is there. One clue is a sheet of cloud or haze through which mountains poke like islands. The top of the sheet often lies at the bottom of the inversion. If the inversion slowly rises or sinks past a mountain top there can be dramatic changes of temperature there. The two main ways that an inversion aloft forms are:

   (a) air below is cooled - say when it flows from low latitudes across a cool sea, as in the south-west winds of a warm sector (map 11A) or the east winds of a summer high over Scandinavia (map 7A, where Cairn Gorm summit is little cooler than winds from the North Sea along the east coast);

(b) air above is warmed - as by slowly sinking in a high or ridge of high pressure.

3. On sun-heated slopes when there is little wind, warm gusts rise to the tops (diagram 20C).

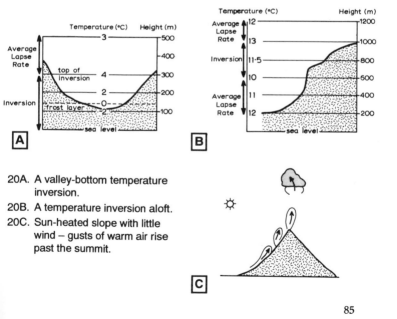

20A. A valley-bottom temperature inversion.

20B. A temperature inversion aloft.

20C. Sun-heated slope with little wind – gusts of warm air rise past the summit.

21. Dull Top

Many clouds over Britain have bases below 1,200m, so it is no wonder that our mountain tops are often in or near cloud, the more so in winter. But days when the forecast speaks of broken or even little cloud over open country can be dull and sunless over the hills for hours on end. The reason is that mountains can make their own clouds. As we saw in Sections 5 and 6, most clouds are made by the lifting of moist air - in the tumbling eddies of strong winds, or the buoyant masses of warm air on sunny days, or the widespread gentle lifting near lows and fronts. But even if clouds are not forming by these kinds of lifting, other clouds can form when air rises over a mountain barrier.

Air rising up the windward side of a ridge can give cloud if it is moist enough. You can watch clouds forming there in more or less the same place for hours on end. They start at the base as small shreds, rise, grow, and join to form a mass that seems to sit on or over the highest ground (colour photograph 21A). On the leeward side, where the air is sinking, the cloud breaks up because it is being compressed as it sinks (diagram 21B). (An example of air being warmed as it is compressed is the pumping up of a cycle tyre.) The cloud may be just a small cap on an isolated hill (diagram 21C and colour photograph 21D), or a vast grey shroud of stratus or stratocumulus over a broad highland, beneath which the deep valleys run like gloomy tunnels (diagram 21E).

Watch out for dull tops whenever a moist wind is forecast to blow over the mountains - as when the weather map shows the wind has come across the sea from low latitudes (maps 5B, 12F, and 15E), or there has been rain for several hours (maps 3D, 14C and 27A). Expect the cloud base to rise, even to above the tops, if the wind becomes drier, as it may well do if a fitful sun makes the day warmer, or if the wind direction changes to blow from higher latitudes (say, after a cold front has passed - maps 3E, 12F, and 15E), or from a warmer country in summer.

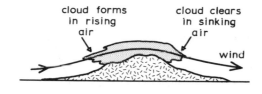

B

21B. Cloud continuously forms and clears in the same places as moist air crosses high ground.

C

21C. Wave-like cloud cap on an isolated hill.

E

21E. Vast cloud mass over highlands, with deep valleys turned into tunnels.

22. Cloudy Top

Apart from wholly sunless days, the tops can be cloudy with only fitful sun even when the forecast is for 'sunny intervals' over low country. On a day when there are cumulus clouds, those carried on to windward slopes can grow more strongly in the air forced to rise over high ground, or they may even be set off there. Either way, cumulus clouds, and any stratocumulus formed by the spreading of their tops, tend to be deeper and longer lived over mountains than over low ground, with the tops passing in and out of cloud (diagram 22A)

Even on days with little or no wind, cumulus clouds tend to form first and grow larger over mountains compared with low ground. There are several reasons for this. After a clear, quiet night there may be a ground inversion through which the mountain tops poke. Cumulus clouds cannot start over the lower slopes until the inversion has been destroyed by the sun's heating, but over the higher slopes they are not thwarted in that way (diagram 22B). Moreover, sun-facing slopes warm up faster than flat ground, so masses of warm air can break away from there and rise to give cumulus clouds (diagram 22C). The positions of individual cumulus clouds may be tied to patches of vegetation, rocks and snow, because these warm up at different rates.

Remember, the base of cumulus clouds usually goes up during the daytime, but it may well come down again in the late afternoon on windward coasts, or indeed at any time of the day if the wind brings in moister air.

A *banner cloud* (diagram 22D) can form in the wake of an isolated mountain around which the wind blows freely, but where a stationary eddy raises air further than in any eddies there may be moving on the wind. Cloud appears at the top of the eddy where it turns downwind, its edges being ragged because it mixes with surrounding cloudless air.

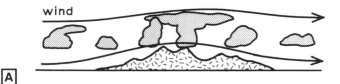

A

22A. Cumulus clouds tend to be deeper and longer-lived over mountains than over low ground.

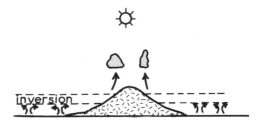

B

22B. Ground-heated air does not rise far enough beneath the inversion to form clouds, but it does over the higher slopes.

C

22C. Cumulus clouds forming over sun-heated slopes.

D

22D. Banner cloud in the lee of an isolated mountain.

23. Cloudless Top

On some days with a general forecast of dull, sunless weather, you can be on the tops in hours of bright sunshine. There is hardly a cloud overhead, yet a 'sea of clouds' below. The highest peaks and ridges come through the soft, quilt-like sheet as though they were islands, and the great clarity of the air lets you see far into the distance (diagram 23A).

This kind of weather is not common but it can happen when a large high sits over Britain and all the clouds lie beneath an inversion aloft that keeps below the mountain tops. Winter is perhaps the most likely time of year to find this weather because the heat of the summer sun tends to keep the inversion too high, except perhaps near windward coasts. At sun-facing slopes on windless days, the edges of the cloud sheet can be seen breaking up, and shreds are carried upwards in air rising from the warm ground (diagram 23B). If you stand with your back to the sun and look down on the cloud you may see a *Brocken Spectre* - your own shadow stretched out over the cloud top. Around the shadow of your head may be coloured rings known as a *glory*, but you cannot see your companion's glory!

Seen from a valley, the grey sky does not look promising, but if it is not too dark, the weather map shows a high, and there is no drizzle despite a low cloud base, suspect that the cloud is shallow and the mountain tops clear, the more so if there is a gap through which no higher clouds can be seen. The discomfort of climbing through cloud may be rewarded by the bright, sunny tops. Even if the tops are still just in cloud they can clear later if the inversion sinks a little. On the other hand, watch the cloud top below you to see that it does not rise and swallow you up unawares.

Sometimes a sheet of very low cloud does not reach the top of a ridge because the wind is blocked (diagram 23C). The leeward side is then free of this cloud (although there may be others at greater heights).

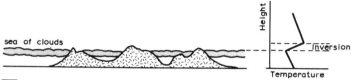

A

23A. Sea of clouds, with mountain tops like islands above the temperature inversion.

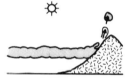

B

23B. Shreds rising from the edge of a sea of clouds where it lies against a sun-heated slope.

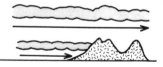

C

23C. Lower cloud sheet blocked by mountains, giving a cloud-free lee side.

24. Hazy Top

Days with widespread haze may, or may not, bring hazy weather to the tops. If the haze is shallow, as it is more likely to be in winter than in summer, the highest ground may be above the haze top (diagram 24A). But if the haze is deep even the highest tops can be enveloped. In summer, a shallow haze at dawn can deepen to cover the tops by mid-afternoon (diagram 24B). A change in wind direction can bring in a deeper haze from far away.

Haze is thickest just below cloud base. When climbing on a hazy day, the view can become poorer as you get closer than about 300m below cloud base (diagram 24C). This is because the minute particles that make up haze grow into small water droplets by attracting vapour from the air when it is moist enough - as it rises towards cloud base. On very hazy days the base becomes so fuzzy it is not easy to tell when you are in or out of cloud.

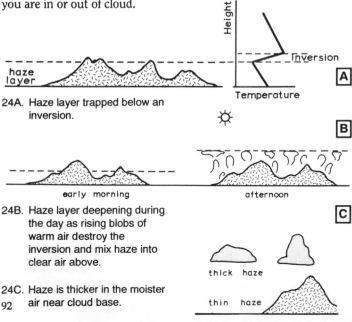

24A. Haze layer trapped below an inversion.

24B. Haze layer deepening during the day as rising blobs of warm air destroy the inversion and mix haze into clear air above.

24C. Haze is thicker in the moister air near cloud base.

92

25. Foggy Top

Clouds enshrouding the tops are often called 'mist'. In them you can see about 100m, but it is rare not to be able to see as little as 50m. This is also called *hill fog*. Clouds down on the hills are common in Britain because the air is often so moist that cloud base is below the tops - a result of winds often blowing from the Atlantic. Hill fog is a hazard to the safe crossing of high ground, so a constant watch should be kept for tops getting covered.

1. If there are already sheets of cloud not far above the highest tops and the wind is changing to become moister, expect the base to come down. Keep an eye on the highest tops, for they are likely to be the first to get covered. Often there is plenty of warning, for the base may fall at a rate of only 100m an hour.

2. If there is a spell of rain, the base may fall more quickly. After a few hours of steady rain, windward slopes may stay covered to below 300m above sea level, until the passage of a front, say, is followed by a drier wind.

3. On a showery day, the air has been moistened patchily, and when a patch flows over high ground, low clouds form quickly but they

25A. Passing patches of moist air formed in showers give short-lived clouds after lifting.

93

may last much less than half an hour. With showers about, expect to see clouds coming and going not only around the highest peaks but also some on the lower hills (diagram 25A).

4. On a cumulus day, when the sky starts blue in the morning, the first clouds are often the lowest, so don't be surprised to see the tops soon cloud over. However, the base may well lift off the tops by the afternoon (diagram 25B).

Hill fog blowing in the wind deposits water on windward sides of rocks, plants and clothes. When the fog is colder than 0°C the deposit is in the form of ice, known as *rime*; it is soft, white and feathery, and over several days can become many centimetres thick. It can harden by partly thawing and then refreezing; it can also become mixed with snow. Rime differs from hard, glassy *black ice*, formed by the freezing of rain or seepage.

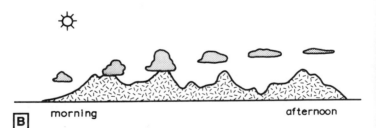

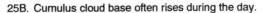

morning   afternoon

B

25B. Cumulus cloud base often rises during the day.

26. Clear Top

Even when the forecast is for widespread haze, coming from faraway places, the tops can stay in clean air so long as the haze is trapped below an inversion aloft. This is most likely to happen in a winter high, where the air above the inversion has been warmed by slow but long-lived sinking. Such air can be so dry there is almost no moisture in it. With long hours of sunshine, and perhaps a wind as well, there is a risk of cracked and sun-burnt skin.

The haze top all around gives a false horizon, and you may be unable to see the sea, which you know to be only a few kilometres away, whereas high tops tens of kilometres away stand out clearly.

Watch for a deepening of the haze layer, for it may swell up to the highest tops. Moreover, when that happens, a cloud sheet may form near the haze top, or spread in from upwind, and the highest ground becomes enshrouded in cloud, so bringing an abrupt end to a gloriously sunny day on the tops (diagram 26).

26. Cloud sheet forms as haze layer deepens when crossing high ground.

27. Rainy Top

British mountains are undoubtedly more rainy than low ground. This is not just because rains are more common there; they also tend to be heavier. Even so, most mountain clouds give no rain.

A forecast of 'occasional rain or drizzle' for low ground can be a soaker in the hills. Most heavy falls over British mountains come with strong south-west winds ahead of a cold front (map 27A). The winds are warm and moist from low latitudes and they are rising gently to give widespread sheets of stratus or stratocumulus clouds, say 1,500m deep - enough to give outbreaks of light rain or drizzle. As the winds rise over the mountains, the clouds deepen, thicken and darken (diagram 27B). If there is not enough time for new rain drops to form (something like an hour is needed) before the wind has crossed the mountains, there will be little rain. But if there are rain or drizzle drops already in the clouds, as they come up to the windward slopes, they grow quickly in the wetter hill cloud to give a drenching rain that goes on and on until the cold front passes, when the south-west winds are replaced by drier west or north-west winds. Such a drenching rain comes on steadily as the clouds deepen ahead of the front, by both base lowering and top rising, and may last five or ten hours if the cold front is slow-moving. The fall can be 50mm to 100mm (2-4in) or more, and there may be floods. Such falls can bring five or ten times as much rain as on low ground to windward. Map 5B shows an example of moist air streaming for hours over mountains to give heavy falls. Some of the drops blow over to the leeward side if the highlands are not too broad.

Rain drops falling from higher clouds near fronts and lows also grow in the thickened hill cloud, leading to heavier rain than over low ground (diagram 27C). Much of the increased rain over hills (even small ones) owes to this mechanism of drops from a 'seeder' cloud aloft washing out water from a 'feeder' cloud below.

27A. Typical weather map with strong, moist south-west winds giving heavy and long-lasting mountain rain. The cold front is moving slowly but steadily south-eastwards. Mild south and south-west winds over England and Wales are being replaced by cooler south-west and west winds over Ireland and Scotland.

27B. Clouds already on the wind deepen, thicken and darken over the hills to give more rain than upwind.

wind

27C. Rain falling from higher, unseen clouds (seeder), and growing in thickened hill cloud below (feeder).

97

28. Showery Top

A day with forecast 'light and scattered showers' can be a wet one in the hills because the showers turn out to be heavy and frequent. We have seen (Section 12) that showers fall from deep cumulus clouds, or the cumulonimbus that grow out of them, and that such clouds tend to be deeper and more frequent over mountains than over low ground (Section 22). It is no wonder, then, that mountain showers can be heavier and more frequent than over low ground (diagram 28A). Indeed, on days when showers are only just able to grow inside the clouds they are more likely to start over the mountains than over low ground. Satellite picture 28B shows a hot summer day with light winds and a high over the North Sea. Showers have already formed over the Grampians, Southern Uplands, Cheviots and Lake District. Other mountain cumulus clouds have not grown as much - for example, Wales, Pennines and North-west Highlands. Note the sunny coasts.

Frequent light showers, each not amounting to much, can plague the hills when low ground stays dry (diagram 28C). What is more, the showeriness is likely to be greater over broad highlands than narrow ridges because the wind will often carry shower clouds well away from isolated peaks during the time it takes for rain drops to grow (diagram 28D). Expect greater showeriness in highland areas, the more so on their windward sides, when showers are forecast to be widespread; and expect showers to go on longer there when they have died away over low ground.

28A. Mountain showers can be heavier and more frequent than over low ground.

28B. A hot summer afternoon with light winds. Showers have already
formed over the Grampians, Southern Uplands, Cheviots and Lake
District. Mountain cumulus clouds elsewhere are less well
developed - Wales, Pennines and North-west Highlands. Note the
cloud-free coasts, and the widespread fog over the North Sea.

Heavy showers tend to be squally because the falling drops set up strong downdraughts that strike the ground and spread out in gusts. The leading edge of a squall is often added to the wind that is already blowing, which can therefore become strong suddenly. Beware of heavy showers upwind. However, a squall at the rear edge may be against the wind, and the combination leads to a calmer spell after the shower.

C

28C. Light showers can fall over the hills when low ground is dry.

D

28D. Showers need not be heaviest on the windward side of highlands because the rain drops take time to grow as the cloud drifts on the wind.

29. Dry Top

It is rare for the tops to be dry when widespread rain or drizzle is forecast. It can happen if the mountains are tall enough to poke through the sheet of rain cloud. But we have seen that clouds need to be at least several hundred metres deep to give even light drizzle, so British mountains are unlikely to poke through most rain clouds (diagram 29A). Such weather is difficult to forecast because the cloud top will be close to the summits of the highest mountains. Moreover, the top is likely to have vast, gentle waves a few hundred metres high that move by, enshrouding the summits from time to time (diagram 29B). Warm sectors are perhaps the most likely places to look for this kind of weather.

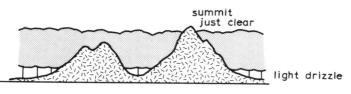

A

29A. British mountain tops are unlikely to poke through clouds deep enough to give rain.

B

29B. Passing waves in a sea of clouds cause summits to go in and out of cloud.

30. Thundery Top

British mountains are not much more thundery than low ground. Any difference may be owing more to distance from the coast than height above sea level. Even so, we should expect more thunderstorms over broad highlands compared with low ground because, as we have seen, they grow from vigorous showers, and showers tend to be heavier and more frequent over broad highlands. This is likely to be true on summer days when storms are first appearing after a warm spell, say ahead of a cold front coming slowly eastwards from the Atlantic (map 14B), or on the edge of a low spreading north from France (map 14C). Torrential rain, even hail, in these storms can give flash floods, and any washing away of a hillside is said to be caused by a 'cloud burst'.

On a ridge or summit in thundery weather you may hear crackling noises and feel your hair stand on end as though it were being pulled. It may even glow with 'St Elmo's Fire'. Retreat quickly to a less exposed place where the risk of a nearby lightning strike is less, particularly if there is hail. Avoid isolated trees. Crouch to reduce exposure, but do not lie down or touch the ground with your hands (to prevent lightning passing through your heart).

31. Snowy Top

When rain is forecast, the more so in winter, we need to judge the chances of *snow on high ground*. Much of the rain that falls at sea level at fronts and in lows starts high in the air as snow. Flakes may start 5,000-10,000m above sea level, take an hour or two to fall, and melt to rain drops only after reaching below 1,000m or 2,000m above sea level. Flakes take a few minutes to melt as they fall into warmer air, so the *melting layer* is often up to 300m deep. In this layer, the partly melted flakes, or flakes and rain drops together, are called *sleet*.

When there is widespread rain inside cloud much of the melting

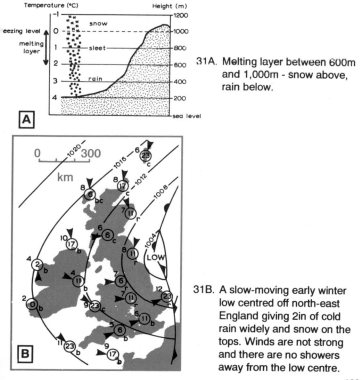

31A. Melting layer between 600m and 1,000m - snow above, rain below.

31B. A slow-moving early winter low centred off north-east England giving 2in of cold rain widely and snow on the tops. Winds are not strong and there are no showers away from the low centre.

31C. Satellite picture for about the time of map 31B. L marks the low
centre. There is widespread rain over northern England and
southern Scotland, with snow on the tops. (This is a night-time,
infra-red image.)

layer has a temperature about 0°C and its base is likely to be no warmer than about 2°C, for sleet in cloudy air is seldom seen at higher temperatures. So, as an example, if the temperature in a valley at 200m above sea level where it is raining is 4°C, and if up to the base of the melting layer there is the usual lapse rate of 1°C for every 200m, sleet or wet snow can be expected above about 600m, and dry snow above about 1,000m (diagram 31A).

As winter approaches, and the average height of the melting layer comes down, widespread snow on the tops can be expected around the cold sides of lows crossing Britain. Map 31B shows a November low slow-moving just off the coast of north-east England, where over 50mm (2in) of rain in two days fell widely over northern England and southern and eastern Scotland. There was over 150mm (6in) in one day in the Lake District, and widespread snow over the tops. The satellite picture (31C) shows the vast cloud mass, although Ireland is almost cloud-free.

Strong winds around lows cause fresh lying snow to drift, so blizzards on the tops can be expected when only rain or sleet falls in the valleys. The records from Cairn Gorm summit in maps 16H and 31D show that long-lasting blizzards were blowing there (with wind-chill making it feel bitterly cold, even if there had been no snow). With long-lasting snow at sea level, the weather can be extreme on the tops. On Cairn Gorm summit, the cold frontal snow of map 15C had a temperature of -7°C and a wind of 29mph - blizzard and severe wind-chill together. Away from lows, even showers with strong winds can lead to blizzards on the tops - for example, maps 3E, 15A and 15E (again, wind-chill making it feel bitterly cold). In the last example, there was a strong gale on the summit making crawling difficult, even if there had not been dense clouds of blowing snow and the risk of exposed flesh freezing.

Snow can fall on the highest tops in any month, although it is of course unusual in mid-summer. At that time of year, brief snow showers may fall on the tops in cold north winds, but there can be more long-lasting snow on the cold side of a low. Map 31D shows a low moving north-east across Scotland in June. There was much rain at sea level but with temperatures there around 7°C it is not too surprising that there was snow on the tops. In fact, as the map shows, the temperature was -3°C at Cairn Gorm summit, and snow settled as low as 250m: snowploughs were needed to clear 6in on the A95 at Glenshee. Snow was covering central Scotland above 900m over the next two days. The satellite picture (31E) for the same time shows the cloud mass around the centre and the cold front trailing over south-east England. The cloud pattern is similar to 31C.

Because the whole depth of the melting layer has a temperature just a little higher than 0°C, as the wind rises over a mountain the resulting cooling can cause the freezing level to fall rapidly - by several hundred metres to near the previous base of the melting layer. This can lead to unexpected icing on the tops.

Snow reaching the ground ahead of a winter front or low often changes to rain after a few hours as warmer air spreads in (see map 15I). Sometimes, however, the opposite happens - rain can change to snow. With long-lasting rain in light winds, the melting layer can build downwards as the melting snow and sleet aloft cool the air. In this way the snow can reach nearer and nearer sea level. A sign that this might be happening is a slow but steady fall of temperature towards 0°C as the rain goes on. In time, a few flecks of sleet can be seen, then more

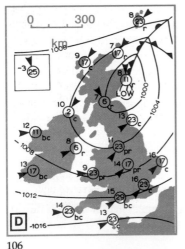

31D. A June low over Scotland giving snow down to 250m in the Cairngorms (inset shows Cairn Gorm summit). Heavy, cold rain at sea level - up to 2in in the Highlands. Some showers in the westerlies over England and Wales.

31E. Satellite picture for about the time of map 31D. Heavy snow over
the Cairngorms in June. 6in lying on the A95 at Glenshee and
snowploughs out. There are only scattered showers over Ireland,
Wales and the Midlands. The cloud-free area over the North Sea
shows air sinking to the east of the low centre.

and more until there are only snow flakes, wet because they have melted a little and clinging to windward surfaces because they are sticky. The bottom of a melting layer has been known to fall 500m in this way over several hours, the more so in heavy rain, where the snow flakes aloft are large and can fall further before melting completely.

Rain sometimes changes to sleet or snow even at low levels behind a well-marked cold front: after the wind veer and as the temperature falls. Tops then become covered quickly. Map 3E is an example.

If cloud base is below the melting layer, settled snow on the higher slopes cannot be seen from the valleys, so there is no obvious sign of the bad weather on the tops. But up there, with snow falling in cloud on a snow-covered ground, it can be a *white-out*, where shadows disappear, avoidance of snow-filled hollows becomes difficult, and horizon judging is very hazardous, especially near the tops of crags.

If cloud base is above the melting layer, not only may settled snow be seen on the higher slopes, but falling snow flakes are chilled somewhat by partial evaporation into the somewhat drier air so the melting layer falls even further. Wet snow can freeze after settling even if the temperature is a little above 0°C, so long as there is a dry wind.

32. Windy Valley

Winds in valleys can play strange tricks. The general forecast may speak of moderate speeds yet there can be strong or gale force winds in one valley but not in another nearby. There are many reasons why valley winds can be stronger than over open country.

1. When the wind can hardly rise over a ridge it tends to blow strongly through any gap there may be (diagram 32A). Some mountain passes are named after their strong winds. On the leeward side the wind can shoot out as a strong jet like water through a sluice, or fan out into a moderate wind. A similar jet can form downwind of a constriction in a valley.

2. When the wind blows into a narrowing valley, there may be a strengthening up to, and more so over, the pass at the end of the valley (diagram 32B). Such a *funnelling* is common in deep valleys lying more or less along the wind. For example, a valley lying E-W is likely to have a west wind if the weather map shows a wind from between SW and NW, but an east wind if the map shows NE to SE. But if the wind at valley bottom becomes separated from that aloft there can be an even greater difference in direction: the wind blows along the bottom from high to low pressure.

32A. Wind strengthening through a gap in a ridge, causing a jet on the lee side.

32B. Funnelling of wind along a narrowing valley.

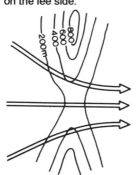

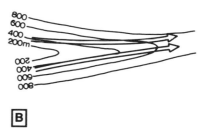

3. In showery weather, downdraught squalls (Section 28) may be channelled down slopes and along valleys, leading to sudden strong winds even on days when only light to moderate winds are forecast.

4. Along coasts on quiet, sunny days, an onshore wind is likely to start up during the morning once the land becomes warmer than the sea. This *sea breeze* may reach force 3 or 4. It can be channelled up valleys, where its onset may be sudden. Coming from over the sea it is cool, and being up to 1,000m deep it can flood into most British highlands, passing through gaps but perhaps being blocked by the highest ridges (diagram 32C). After sunset this wind dies away. Some large lakes have coastal breezes but they are likely to be weak.

5. Sometimes surprisingly strong winds blow *across*-valley. This happens when air flows over an upwind ridge, like water over a weir. The strengthening wind down the lee slope is known in the Eden valley, Cumbria, as a *helm wind* - a term that may be used more widely for this kind of wind. (Another term is *fall wind*.) Watch out for such a strong wind whenever an inversion aloft lies just above the ridge line (diagram 32D), more so at night than by day, and in dull weather rather than sunny. Speeds can reach gale force although only moderate to fresh winds are forecast, and they may become violent when gales are widespread. In sunny weather, the helm wind is likely to be intermittent, with sudden strong downdraughts every few minutes. If the wind descends into a

32C. Sea breeze approaching mountains.

32D. Strong lee-slope wind (helm wind) when a temperature inversion aloft lies not far above ridge height.

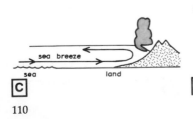

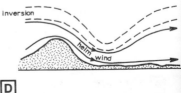

restricted hollow it may be deflected by the walls. Even so, remember that a ridge seldom has a straight top; some parts may poke above the inversion and any helm wind will blow only in the lee of cols. Moreover, because the inversion may rise or fall during the day, as the weather system changes, there may be changes in the extent and strength of the helm wind. Don't wonder at a sudden across-valley wind springing up when you thought, after looking at the weather map, you would be in a sheltered place.

6. A wind blowing around a hill is strengthened on the shoulders, sometimes to as much as twice the speed over open country (diagram 32E). This kind of strengthening is like the flow around a boulder (as shown by scoops in lying snow - photograph 16D) but on a much larger scale. It is most likely when there is an inversion aloft lying below summit height, for there may not be enough energy in the wind to lift the inversion above the summit height.

7. Winds in the wake of a hill, rather than a ridge, can be very complex because the flow may be over or around. Compare with winds near a building, or river flow near an isolated boulder. There may be organised eddies, both flat and upright, that either stay in place or are shed downwind from time to time. Gusts make the wind erratic and there may even be small, short-lived whirlwinds, revealed by lake spray or blowing snow. It is probably simplest to say that both speed and direction of the wind will be highly variable for a long way from the hill! The ratio of this distance to width of hill is like that of long snowdrifts in the lee of a boulder.

It is very difficult to forecast just where and when these strong valley winds will be blowing. Expect them whenever the general forecast speaks of winds of more than moderate strength, but use your topographic map and the general weather forecast to judge where strong valley winds are likely to blow.

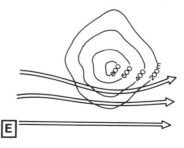

32E. Wind strengthening as it blows around the side of a hill.

33. Calm Valley

On days when strong winds are forecast, some valleys can still be calm. The most likely ones are those lying across wind but, as we have seen (Section 32), even there the wind can be remarkably strong. On many days when there is at least some sun the wind blowing across a ridge leaves the ground near the ridge line or crag top, and the leeward valley is more or less filled with a slowly overturning eddy (diagram 33A). The valley bottom wind is then not only light, it blows against the wind direction expected from the forecast. It contrasts strongly with the wind at ridge height, as is shown up by racing clouds or their shadows on the slopes. These eddies, with their light winds, can give way at times to a helm wind, so a valley that might be thought to be sheltered turns out to have surprisingly variable winds, some strong enough to be uncomfortable if not hazardous. Similar but smaller eddies can form at the foot of the upwind slope of a ridge or a crag.

Even when a more or less steady wind blows along a valley there can be calm places, such as behind a knoll (diagram 33B), a spur on the valley side, or just beyond a bend (diagram 33C). In all these places there can be eddies; the first has a level axis, the others more upright. But all these eddies are likely to come and go: at times the wind spills over the knoll or around the spur or shoulder.

On sunny days when little or no wind is forecast (often within a high), a gentle breeze (up to force 2) can start to blow up-valley during

33A. Slowly-overturning eddy in a valley lying across the wind.

33B. Calm patch in the lee of a valley-bottom knoll.

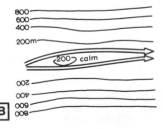

112

the morning, dying away towards or soon after sunset. Such *up-valley winds* can bring some relief on an otherwise hot, still day. They are driven by the sun heating a smaller volume of air in the valley than over open country with the same area, but they have not been well described for British mountains. There may also be *up-slope* (or *anabatic*) *winds* on sun-heated slopes, which the up-valley wind can feed by drawing in air from open country (diagram 33D). Because of their different aspects, opposite slopes respond differently, with an anabatic wind appearing on one but not the other and leading to a gentle across-valley wind. For example, in a north-south valley the slope facing west has a later sunrise than the one facing east, and changing the axis direction also changes the timing of onset.

An up-valley wind can spill over the pass at the valley head, continue down the next valley and so undercut any up-valley wind there.

On clear, quiet nights there may be a *down-valley wind* (again up to force 2), driven by loss of heat to space from a smaller volume of air in the valley than over open country with the same area. It blows more or less in line with valley walls but there can be meandering, like a river, and bouncing off the walls at bends. It is also fed by cool air draining down the valley sides as shallow and gentle *down-slope* (or *katabatic*) *winds* that start soon after each slope enters shadow later in the day,

33C. Eddy just beyond a sharp bend in a valley.

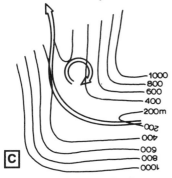

33D. Sun-heated slopes cause up-slope (anabatic) winds fed by an up-valley wind.

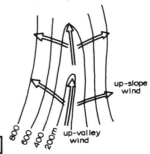

the more so over higher slopes more open to the sky or where there is a cover of snow. Surprisingly gentle slopes can be effective: even as little as 1:100. Such weak winds may be shown up by the creeping of valley fog.

Down-valley winds tend to come in pulses - cool 'air avalanches' at intervals of 10 or 20 minutes (or even an hour or more) with calms between. Although these winds are weak, and are easily modified by changes in slope or rock outcrops or stands of trees, they can be significant in deciding where to camp. Where valleys meet, the colder air from one may undercut the less cold from another, so in broken country the pattern of valley and slope winds can be complex. Where a valley leads into open country, the down-valley wind forms a jet flowing out some kilometres, as may be seen in a ribbon of fog taken out to sea (satellite picture 40A).

Down-slope winds can appear over snow or ice even during the day. In fact, such *glacier winds* tend to be strongest in the afternoon, when the temperature contrast is greatest, and they blow down to snow-free slopes.

Because both up- and down-valley winds are weak they are easily swamped by only moderate winds building down from aloft. If this broad-scale wind increases during the night it scours the top of the down-valley wind and may eventually replace it, accompanied by a rise in temperature. This is more likely to happen if the valley is shallow.

34. Cold Valley

A sheltered valley on a clear, calm night can be colder than open country by 5°C or more. This is partly because any wind over open country stirs the cooling through a deeper layer than in the valley. Cooling in the valley as the wind drops to near calm around sunset causes a very strong ground inversion to form. As cooling goes on through the night, some deep valleys can have a frost by sunrise, even in summer. In winter, with a snow cover, the temperature can fall to below -20°C. When camping, therefore, it would be wise to seek a place 100m or more above valley bottom so as to miss the coolest air. Even a 50m difference can be 5°C warmer (see Section 20), but the inversion can sometimes fill a valley. On such clear, calm nights, temperatures in valley bottoms are much the same no matter what their heights are above sea level. Much the same thing can happen in sheltered hollows high in the hills; they are known as *frost hollows*.

If the large-scale wind strengthens during the night it can steadily erode the top of the inversion until the remainder is flushed out. At the valley bottom, the onset of wind brings a rise in temperature.

After sunrise the inversion starts to be destroyed from below through contact with the warming ground, and also from above because air over the ......dle of the valley sinks slowly (and therefore warms) in response to the up-slope winds. The inversion may be wholly destroyed by around midday, when the up-slope winds tend to be replaced by separate 'thermals' of warm, buoyant air (exploited by glider pilots to gain lift). Even with a snow cover, weak up-slope winds may still appear if there are bare coniferous forests warmed by sunshine, so heating from above is still possible even if heating from below is small.

In a steep-sided valley the sun may not be able to shine on the bottom, the more so in winter and when the valley lies east-west, not north-south, say. Such a valley can stay colder by day than open country, as shown by hoar frost lasting on shaded ground. Moreover, snow lies longer and helps to keep the air cooler. If the cool air is unable to drain away easily, winds stay light and more or less cut off from winds at greater heights. The cold air in the valley bottom behaves something like water in a bath - it can surge gently backwards and forwards, or from side to side, so that on the hill slope cold air wells up and down more or less rhythmically, with periods of minutes or hours, depending in part on valley size (diagram 34A).

After a warm day with an up-valley wind, a down-valley wind may set in suddenly. At onset, as the wind direction reverses, the temperature can fall several degrees in as many minutes (diagram 34B). But at some places, where the down-valley wind starts late and after the daytime wind has dropped to calm, and a ground inversion has started to form, as the wind picks up there is a temperature rise.

34A. Valley-bottom temperature inversion surging to and fro, causing the temperature to rise and fall on the slopes.

34B. Down-slope (katabatic) winds on the shaded, east-facing slope have started before up-slope winds on the opposite slope have stopped. A cold, down-valley wind suddenly replaces a warm up-valley wind.

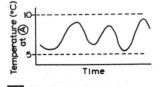

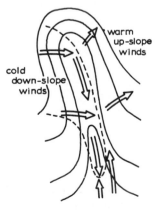

A

B

35. Warm Valley

On some days a valley can be warmer than open country. There are four main ways this can happen.

1. On a sunny day, a valley floor becomes hotter than open country at the same height. This is because not only does the sun heat a smaller volume of air in the valley than over open country with the same area but also the cooling by loss of heat into space is slower in the valley, for the walls make the sky there smaller than in open country (diagram 35A). Moreover, the walls radiate heat back to the floor. In a north-south valley, the west-facing wall is the warmer of the two at the time of highest temperature in the afternoon.

2. In a rain shadow (Section 44), where the wind is blowing down from rainy hills, the air is warmer than at the same height on the windward side (diagram 35B). It became warmer as it lost some of its moisture as rain. The warm wind on the leeward side is called a *föhn wind*. It is warmest in valley bottoms. On the other hand, a wind crossing the hills but giving no rain is not warmed in this way.

35A. Rate of loss of radiant heat to space from the valley bottom is less than from open country; hence the valley becomes hotter by day.

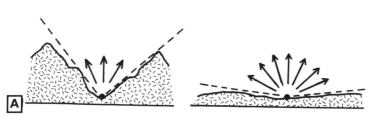

3. When an inversion aloft lies below ridge height, winds below it may be unable to cross the ridge - they are blocked - so the wind sweeping down the leeward slope has come from above the inversion and it is warmed by compression as it sinks (diagram 35C). This warm wind is another kind of föhn, found not when the hills are rainy but even when the tops are cloud-free, above the inversion (Section 23). (If the inversion were to rise above ridge height, this warm, föhn wind may be replaced by a cool, helm wind - Section 32).

4. If there is a föhn wind of either kind, the valley is likely to have broken cloud, so sunshine there gives warmer weather than on the cloudy, windward side of the hills.

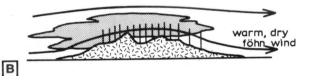

B

35B. A wind blowing down from rainy hills is warmer than on the windward side at the same height. It is one kind of föhn wind.

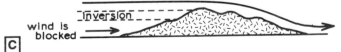

C

35C. Warm wind blowing down a lee slope from above a temperature inversion when the wind beneath is blocked. It is another kind of föhn wind.

36. Dull Valley

Days when widespread broken clouds are forecast can be dull and sunless not only over the hills but also in the valleys running among them. By and large, we should think of a wind blowing over a highland area in one sweep and not much affected by valleys lying across the wind. Sheets of cloud formed by the lifting will then cover hill and dale alike. Slowly overturning eddies in cross valleys more or less stop the overriding air from coming down into the valleys (diagram 36A). Cumulus clouds do much the same thing as they drift by in the wind, but because on a given day some hills are better than others at setting off or enlarging cumulus clouds, parts of a valley can stay sunless whereas other parts get fitful sun.

Sheets of stratus clouds over low country, capped by an inversion aloft, can be blocked as they are brought up to a ridge on the wind. Only where a valley lies into the wind can the low cloud spread among the hills. Indeed, such an exposed valley can stay dull and cool whereas another one nearby, lying across the wind, is sunny and warm with a föhn wind (diagram 36B). Such differences between valleys are similar to those where fog is spreading inland from the sea (see satellite picture 40B). Cool, moist air feeding up a valley, after being heated over the hills, leads to cumulus clouds that are more likely on the windward slopes (and have a lower base) than around valleys not reached by the moist air.

A

36A. Valley eddies causing the wind to blow across highlands in more or less one sweep.

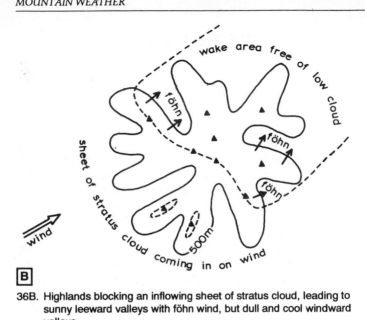

B

36B. Highlands blocking an inflowing sheet of stratus cloud, leading to sunny leeward valleys with föhn wind, but dull and cool windward valleys.

37. Cloudy Valley

Apart from more or less broken cumulus and stratocumulus clouds, which we have seen are common over British mountains (Section 6) and, by drifting on the wind, also over nearby valleys, there are other kinds of broken cloud found mostly over valleys.

1. When there is an eddy in a cross valley, and rain has been falling for hours so the valley air is very moist, then shreds of cloud can form in the upward-moving part of the eddy. These clouds are best seen where there is a dripping forest on the hillside, and very moist air is drawn fitfully from among the tees (diagram 37A). Such shreds can be the first of much more widespread low clouds to form in a long spell of rain.

2. When a cold front passes overhead and the wind changes direction from along a valley to cross it, then some low clouds from the warm airstream become trapped as the eddy forms. Such low clouds show up against the rising base of higher clouds behind the front (diagram 37B and colour photograph 37C). These 'left-overs' go at last by mixing slowly with overriding air, or because the ground is sun-heated during he daytime.

3. When there is an inversion aloft, the first cumulus clouds to form during the morning can be over the lower slopes beneath the

37A. Moist eddy in a cross-wind valley causes cloud shreds to climb slowly up the windward side.

37B. For a time after a cold front has passed, patches of low cloud from the earlier warm air may linger in the eddies.

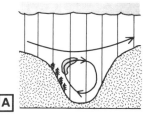

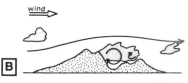

☼

D

37D. The first morning cumulus clouds can be over the lower slopes when there is a temperature inversion over the upper slopes.

E

37E. Wave clouds forming in the crests of air waves over and in the lee of a ridge.

inversion, and the mountain tops stay clear (diagram 37D). Later, the sun's heat may destroy the inversion and cumulus clouds form over the tops as well.

4. When the wind blows across a ridge and sinks back on the leeward side it can overshoot and set up vertical oscillations that are stretched out as a train of waves, each typically 5-30km apart. These *lee waves* move through the air with a speed equal to but opposite the wind, so they are stationary relative to the ridge. In their crests smooth, lens-shaped or banded clouds can from (diagram 37E and colour photograph 37F). These *wave clouds* (or lenticular clouds) are more or less fixed in the sky and the wind blows through them, as a river flows through standing waves downstream of an underwater boulder. Where there are many ridges with a great variety of heights and shapes, these clouds can be scattered over the sky, or in lines, or even stacked one above the other ('pile of plates'). Waves from adjacent mountains can

37G. Waves with a rotor, giving a patch of winds blowing against the main stream.

37H. Wave gap in the trough of a lee wave.

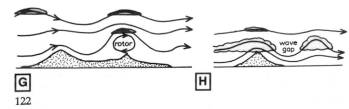

G

H

37J. Wave clouds are widespread from northern Scotland to southern England, in a sheet of stratocumulus with a top near 1,500m. Spacing is around 10km, and shadows are to the north-west because this is a morning picture. There is fitful sunshine under the gaps between the wave clouds. South-west Scotland is cloud-free because the wind blows down in the lee of the Highlands. In contrast, the cloud sheet passes less disturbed down the east coast. The 'pancake' clouds over the North Sea are the spreading tops of cumulus giving light showers.

123

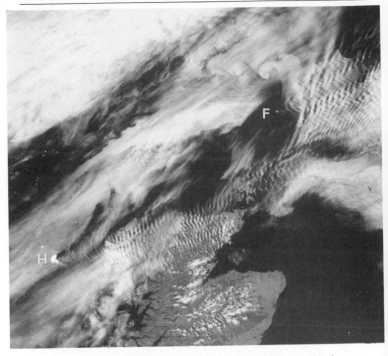

37K. Stratocumulus clouds over the sea on a hot and sunny day in Scotland. V-shaped wakes in the lee of the islands Hirta (H) and Foula (F) show the wind is west-south-westerly. Other wakes downwind of Shetland cross each other to give a herring-bone pattern.

interfere, leading to complicated cloud patterns. Wave clouds can be a forerunner of cloudy or even wet weather, for they show there are sheets of moist air that may be lifted later to form clouds more widely as a front or low comes nearer. Beneath some waves the wind can turn right over as a *rotor* (diagram 37G). In the bottom of a rotor there can be strong gusts, blowing against the main wind stream, and they can be dangerous if they happen to lie over hill tops.

5. If there is already a sheet of cloud in a wind that is giving waves in the lee of a ridge, there can be gaps in the troughs of the waves. These *wave gaps*, too, are more or less fixed in the sky; hence places beneath them get more sun than places only a kilometre away (diagram 37H and colour photograph 37I). Satellite picture 37J shows wave clouds and wave gaps in a vast sheet of stratocumulus cloud that extends from northern Scotland to southern England in north winds. There are widespread waves, but south-west Scotland is cloud free because it is in the lee of the Highlands. For more examples of trains of wave clouds, see satellite pictures 5D (widely over Scotland, in stratocumulus based around 1,200m), 6I (starting over the Lake District and Pennines, in stratocumulus based around 1,000m), 15B (Connemara and Scotland), 15D (Wales) and 31C (starting over the Cuillin).

6. On rare days, wave gaps in the lee of an isolated peak can be in the form of a 'V', like the wake of a moving ship. Such a wake is more easily seen from above than from the ground; satellite picture 37K is an example with moist west-south-west winds over the sea, and stratocumulus clouds in which the islands Hirta (H) and Foula (F), both about 400m high, are forming wakes, with waves about 6km apart. More wakes are to be seen in satellite pictures 5D (in stratus to the lee of Scilly and Isle of Man) and 37J (in stratocumulus to the lee of various peaks in the North-west Highlands of Scotland - trailing into wakes caused by the Grampians - as well as the Wicklow Mountains, Ireland).

38. Cloudless Valley

A day forecast to be dull with an unbroken sheet of stratus cloud can be sunny and more or less cloudless in a sheltered valley. This happens when a ridge blocks the flow of clouds into the valley because there is an inversion aloft. If cloud top is below ridge top, no cloud may be seen from the leeward valley (diagram 38A). But if clouds rise just above ridge top, they may be seen spilling over like waterfalls, endlessly evaporating in the same place. Such a cloud on the ridge is known as a *föhn wall* (diagram 38B). The quilt-like pattern on its upper side can be seen sliding into the downwind edge. Where a ridge varies in height along its length, there may be spillage over the lowest parts but blocking elsewhere (colour photograph 38C). Moreover, if the cloud top wells up or down so the spillage grows and shrinks.

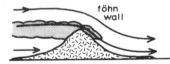

38A. Cloud sheet wholly blocked by a ridge.

38B. Cloud sheet partly blocked - the top pours over and evaporates by compression at the föhn wall.

39. Hazy Valley

A day when clean air sweeps across the country, from far away with few sources of pollution, can still be hazy in the valley if it is sheltered and there are local sources within it. Smoke may come from factories, or the house and rubbish fires of a town. With an inversion aloft, the smoke may be spread up to its base, perhaps 1,000m or 2,000m above the ground. But with a ground inversion the smoke may be taken up no more than 100m, to gather in blue or grey sheets as each plume first rises and then spreads sideways (diagram 39A). After a clear, calm night in winter, the lighting of morning fires can thicken the haze to a pall through which things can be seen up to only a kilometre or two away.

Such a haze can be cleared away if the wind changes direction to blow along the valley. It can be thinned by mixing over a greater depth, as can happen during the daytime, when the sun's heat sets up-slope winds blowing (diagram 39B) and cumulus clouds may form later.

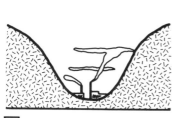

A. 39A. Smoke plumes spreading within a valley-bottom temperature inversion.

B. 39B. Smoke mixes upwards as the sun's heat destroys the temperature inversion.

40. Foggy Valley

On clear, quiet nights when the air is moist, fog is a possibility as the temperature falls (Section 9). It is often only 100m or 200m deep so valleys may be filled whilst even modest hills poke through like islands. Satellite picture 40A shows a winter morning with widespread freezing fog (lasting all day in places and temperature not rising above -3°C), but many hills are clear. Fog from the Severn valley has spread over the estuary, and many smaller valleys in England and Wales have ribbons of fog. After sunrise, large areas of lowland fog like this shrink inwards from the edges (where it is shallowest) as the sun warms the ground steadily and hence the foggy air above.

Fog is fickle. After a clear night with a forecast of fog, valleys can often be fog-free, or else there are only patches. Moisture in the air has condensed out as dew instead. The reason for this is probably that valley winds during the night do not fall light enough to let fog form. In open country, the wind needs to be no stronger than about force 1 for fog to form, whereas down-valley winds on clear nights are often a little more than that, as can be seen in the drift of smoke. Where the down-valley wind drops out to calm, say near barriers such as tree clumps or a narrowing of the valley, fog patches are likely. Where a stronger breeze blows in from a side valley, fog is unlikely.

Valleys can have fog that spreads in from nearby sea or low ground. It might be thought that this would be impossible because the wind ought to mix cold, foggy air with warm, dry air aloft, so that the fog would evaporate. But once widespread fog has formed, the ground inversion builds upwards to near the fog top, and mixing with clear air above becomes more difficult. Even a force 2 breeze can then carry fog. Any lifting needed to take a fog up into a highland valley may well add a little to the cooling of the foggy air. Look out for days when light winds blow from places with widespread fog. Expect a sudden temperature fall as the fog arrives. Satellite picture 40B shows widespread North Sea fog during the morning. Overnight the fog has crept up many valleys in the eastern side of Scotland and north-east England, leaving some hills like islands. See also satellite picture 8.

Fog will not form over land if there is much cloud. Because some

40A. Mid-morning in winter. Many hills poke through a sea of freezing fog
- for example, Cotswolds (C), Chilterns (L), Berkshire Downs (B),
Salisbury Plain (S), Mendips (M), Quantocks (Q) and Forest of
Dean (F). Ribbons of fog fill many valleys. Where valleys reach the
coast, fog goes on out to sea, showing that night-time down-valley
winds are still blowing - for example, valleys of the rivers Exe (E),
Axe (A) and Adur (D).

valleys can be more or less cloudless on an otherwise cloudy day
(Sections 37 and 38), watch out for unforecast fog patches in those
valleys.

Fog drifting over a lake may be lifted from the surface because the
water stays warm at night. However, on some nights a drift of cold air
over a lake or river causes steaming of the warm water surface. Such
steam fog may drift onshore and help to set off fog over land.

40B. A spring morning when North Sea fog had crept up many valleys
overnight - in northern Scotland (for example, Strath Oykell), the
Scottish Lowlands as far west as the Firth of Clyde, and north-east
England (for example, the Tyne valley, Wensleydale and
Wharfedale). Hills stand out like islands (for example, Campsies,
Ochils and North York Moors). Snow lies on the highest tops, for
example, around Ben Nevis, the Cairngorms and the North-west
Highlands. Fog clearance over Orkney has spread downwind. Cities
show as dark patches (for example, London, Birmingham and
Manchester) and so do uplands (for example, Dartmoor, the Welsh
mountains, the Pennines, the Lake District and the Southern
Uplands, as well as the mountains of Wicklow, Kerry, Connemara
and Donegal).

130

41. Clear Valley

One valley can be fog-free when others nearby are fog-filled. Because valley fog can spread from elsewhere, a valley can be free if it is sheltered by high ground from the fog-bearing breeze. This can happen in much the same way as when a ridge blocks a stratus cloud sheet (Section 38). Satellite picture 40B shows how the Pennines, for example, are sheltering north-west England from North Sea fog. If there is a föhn wind, blowing down from above fog on the windward side of the ridge, valley fog is even less likely (diagram 41).

41. Blocked fog sheet and lee-side föhn wind.

42. Rainy Valley

A rainy day in the hills is likely to be a rainy one in the valleys as well because rain drops drift on the wind. Most drops fall at 10-20kph. Because they have to fall at least a kilometre or two through and below the clouds where they started, they take some fraction of an hour to fall out, and in so doing will drift several kilometres on most winds. Most valleys among British hills are no wider than this, so even if the rain cloud sits over the hills, and does not move with the wind, the downwind valley will still get rain, often plenty, but more in the form of small drops that can drift furthest (diagram 42A and colour photograph 42B). This is perhaps most likely in the driving rain ahead of a cold front. When a low, fitful sun shines into the valley a rainbow comes and goes. Snow flakes are carried even more easily on the wind because they fall slower - only a few kilometres an hour. Hence the rain that comes from their melting falls even further downwind.

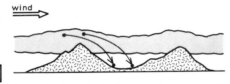

42A. Curved paths of rain drops as they grow whilst drifting into a valley after starting well upwind.

Because some mountains are better able than others to form rain clouds, or to renew clouds already on the wind, it follows that clouds just in their lee will have more rain than those over other valleys. For example, on a day with seemingly endless rain in a strong, moist south-west wind, there can often be seen, looking along a cross-wind valley, curtains of rain sweeping down on some parts more often and more heavily than on others. Likewise, valley heads are more often rainy than valley feet (diagram 42C).

Sometimes there are other, unseen rain or snow clouds at greater heights caused by an approaching low or front, or by waves in winds rising over hills. Rain from these clouds can spread far downwind, even at 50-100kph in the winds at heights of 3,000m to 6,000m (diagram 42D).

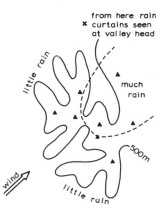

C

42C. Heavier rain over the central highlands can be seen up the valleys.

D

42D. Upper clouds, set off by the mountains, can give rain far downwind.

43. Showery Valley

When showers are forecast, valleys are likely to be as wet as the hills nearby. On the one hand, showers carried on the wind to a highland area can be heavier over the hills than in the valleys because the shower clouds are thickened as the wind rises over the hills. On the other hand, shower clouds that start over the hills are not likely to rain out much until some kilometres downwind, simply because whilst the drops are growing the clouds can drift from the hills. Sometimes the shower cloud itself may not start to form until downwind of a mountain. This can happen on a showery day when the wind splits and blows around both sides of a mountain, and then meets again along a line where the air is forced to rise and so cloud forms there (diagram 43A).

Where the wind takes at least an hour or two to cross a broad highland, showers can lose all their drops before reaching the leeward side. This is why, for example, eastern Scotland and the Welsh Borders have little rain with a showery west wind (diagram 43B and satellite picture 12G). Map 3A shows sheltering around Glasgow with north winds.

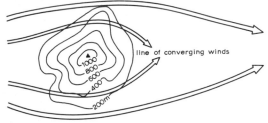

line of converging winds

A

43A. Lee-side line of converging winds where shower clouds are likely to grow.

wind

B

43B. Showers over broad highlands can die away before reaching the lee side.

44. Dry Valley

On a day with rain or drizzle from long-lived stratus or stratocumulus clouds over broad highlands, the leeward side, more so the valleys, can have little or no rain. This happens when the wind is too weak to carry drops across from the windward side where they form (diagram 44A). There can be falls of 10mm or 20mm in the central highlands but little or nothing only 10km away downwind. The dark, gloomy mountains then contrast strongly against the dry, leeward valleys with their fitful sunshine. Looking up a valley towards the hills, not only can you see curtains of rain sweeping along in the wind, but also sometimes the downwind edges of the cloud mass are seen to be breaking endlessly over more or less the same places. Moreover, the base is higher than on the windward side. But if the clouds are deep, only the lowest sheets may break, and some rain may go on falling from the higher sheets (diagram 44B).

The dry leeward side is known as a *rain shadow*. It may not be quite dry, for some drops can come across from time to time (Section 42). Map 4C shows broken cloud and dry weather around Aberdeen when moist, cloudy west winds are giving long-lasting rain and drizzle over western Scotland. Rain from lows and fronts can be much less than expected to the lee of broad highlands. Map 16F shows no snow over west Wales when there was widespread snow over southern Britain.

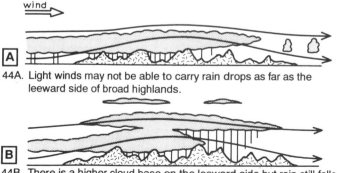

wind

A

44A. Light winds may not be able to carry rain drops as far as the leeward side of broad highlands.

B

44B. There is a higher cloud base on the leeward side but rain still falls there from the upper cloud sheet.

45. Thundery Valley

Just as showers can drift from hills to valleys, so can thunderstorms, because most are really no more than large and heavy showers. On days when thunder is forecast, don't expect the valleys to escape. However, some valleys, or parts of them, may get more thunder than others, on average, if storms are more likely to grow over or near some mountains than others. Storms are sometimes said to 'move in circles', but this would be must unusual. Two or more short-lived storms moving along close but straight paths might give this impression.

46. Snowy Valley

Falling snow in a valley is unusual when only cold rain is forecast for open country at the same height. It can happen when cold air has been trapped in a sheltered valley, say ahead of a warm front in winter when winds are light. Such snow is likely to be short-lived.

REMEMBER

- Mountain weather is fickle, and its changes are difficult to forecast.

- Weather on the summits can be unexpectedly severe, particularly on the higher and more northerly summits of Scotland.

- Before planning a day among the mountains get a forecast and judge if the weather is likely to be favourable.

- Before starting out, get the latest and most detailed forecast. Change your plan if the weather is likely to turn foul.

- Whilst outdoors, take an interest in the weather, and watch out for any marked differences from the forecast. Change your plan if the weather is turning foul.

Be weather wise!

FURTHER READING

Introductions

The following are just a few; there are many more to choose from.

Scotland's Winter Mountains, Martin Moran - David & Charles (1992)

The Story of Weather, Bill Giles - HMSO (1990)

Weather: A modern Guide to Forecasting, Bruce Atkinson and Alan Gadd - Mitchell Beazley (1986)

Weather Watch, Stephen Moss and Paul Simons - BBC Books (1992)

More advanced books

Atmosphere, Weather and Climate, R.G. Barry and R.J. Chorley - Methuen (6th edn. 1992)

Basic Meteorology: A Physical Outline, J.F.R. McIlveen - Van Nostrand Reinhold (UK) (1986)

Mountain Weather and Climate, R.G. Barry - Methuen (2nd edn. 1992)

Weather Systems, L.F. Munk - Cambridge University Press (1989)

CICERONE GUIDES

SOUTH AND SOUTH-WEST LD trails
THE KENNET & AVON WALK
THE SOUTHERN COAST-TO-COAST WALK
SOUTH WEST WAY - A Walker's Guide to the
    Coast Path (2 Volumes)
THE THAMES PATH
THE TWO MOORS WAY
THE WEALDWAY AND THE VANGUARD WAY
SOUTHERN AND SOUTH-EAST ENGLAND
CANAL WALKS Vol 3: South
CHANNEL ISLAND WALKS
WALKING IN THE CHILTERNS
A WALKER'S GUIDE TO THE ISLE OF WIGHT
WALKING IN KENT (2 volumes)
LONDON THEME WALKS
RURAL RIDES No.1: WEST SURREY
RURAL RIDES No.2: EAST SURREY
WALKING IN SUSSEX
THE LEA VALLEY WALK
WALKING IN BUCKINGHAMSHIRE
WALKING IN BEDFORDSHIRE
WALKING IN HAMPSHIRE

SOUTH AND SOUTH-WEST
CORNISH ROCK
WALKING IN CORNWALL
WALKING ON DARTMOOR
WALKING IN DEVON
WALKING IN DORSET
A WALKER'S GUIDE TO THE PUBS OF
    DARTMOOR
EXMOOR AND THE QUANTOCKS
WALKING IN SOMERSET
WALKING IN THE ISLES OF SCILLY

**LONG DISTANCE TRAILS ACROSS
NORTHERN ENGLAND**
WALKING THE CLEVELAND WAY AND THE
    MISSING LINK
THE DALES WAY
THE ISLE OF MAN COASTAL PATH
THE ALTERNATIVE PENNINE WAY
THE PENNINE WAY
LAUGHS ALONG THE PENNINE WAY
THE ALTERNATIVE COAST TO COAST
A NORTHERN COAST TO COAST

LAKE DISTRICT and MORECAMBE BAY
CONISTON COPPER MINES: A Field Guide
CUMBRIA WAY AND ALLERDALE RAMBLE

THE EDEN WAY
THE ISLE OF MAN COASTAL PATH
A LAKE DISTRICT ANGLER'S GUIDE
SHORT WALKS IN LAKELAND
    Book 1: SOUTH LAKELAND
    Book 2: NORTH LAKELAND
    Book 3: WEST LAKELAND
SCRAMBLES IN THE LAKE DISTRICT
MORE SCRAMBLES IN THE LAKE DISTRICT
THE TARNS OF LAKELAND VOL I: WEST
THE TARNS OF LAKELAND VOL 2: EAST
WALKING ROUND THE LAKES
WALKS IN THE SILVERDALE/ARNSIDE AONB
WINTER CLIMBS IN THE LAKE DISTRICT

**NORTH-WEST ENGLAND outside the Lake
    District**
WALKING IN CHESHIRE
FAMILY WALKS IN THE FOREST OF
    BOWLAND
WALKING IN THE FOREST OF BOWLAND
CANAL WALKS, Vol 1: North
LANCASTER CANAL WALKS
A WALKER'S GUIDE TO THE LANCASTER
    CANAL
WALKS FROM THE LEEDS-LIVERPOOL
    CANAL
THE RIBBLE WAY
WALKS IN RIBBLE COUNTRY
WALKING IN LANCASHIRE
WALKS ON THE WEST PENNINE MOORS
WALKS IN LANCASHIRE WITCH COUNTRY

PENNINES AND NORTH-EAST ENGLAND
CANOEISTS' GUIDE TO THE NORTH-EAST
HADRIAN'S WALL Vol 1: The Wall Walk
HADRIAN'S WALL Vol 2: Wall Country Walks
NORTH YORKS MOORS
THE REIVER'S WAY
THE TEESDALE WAY
WALKING IN COUNTY DURHAM
WALKING IN THE NORTH PENNINES
WALKING IN NORTHUMBERLAND
WALKING IN THE SOUTH
WALKING IN THE WOLDS
WALKS IN THE NORTH YORK MOORS 1
WALKS IN THE NORTH YORK MOORS 2
WALKS IN THE YORKSHIRE DALES 1
WALKS IN THE YORKSHIRE DALES 2
WALKS IN THE YORKSHIRE DALES 3

CICERONE GUIDES

WATERFALL WALKS - TEESDALE AND THE
  HIGH PENNINES
THE YORKSHIRE DALES
THE YORKSHIRE DALES ANGLER'S GUIDE

DERBYSHIRE PEAK DISTRICT and EAST MIDLANDS
"Star" FAMILY WALKS IN THE PEAK
  DISTRICT AND SOUTH YORKSHIRE
HIGH PEAK WALKS
WHITE PEAK WALKS Vol 1: THE NORTHERN
  DALES
WHITE PEAK WALKS Vol 2: THE SOUTHERN
  DALES
WHITE PEAK WAY
WEEKEND WALKS IN THE PEAK DISTRICT
WALKING IN SHERWOOD FOREST & THE
  DUKERIES
THE VIKING WAY

WALES, and WELSH BORDER LD Trails
THE LLEYN PENINSULA COASTAL PATH
WALKING OFFA'S DYKE PATH
OWAIN GLYNDWR'S WAY
THE PEMBROKESHIRE COASTAL PATH
SARN HELEN
THE SHROPSHIRE WAY
WALKING DOWN THE WYE
A WELSH COAST TO COAST WALK

WALES, AND WELSH BORDERS
ASCENT OF SNOWDON
ANGLESEY COAST WALKS
THE BRECON BEACONS
CLWYD ROCK
HEREFORD AND THE WYE VALLEY
HILL WALKING IN SNOWDONIA
HILLWALKING IN WALES (2 Volumes)
THE MOUNTAINS OF ENGLAND AND WALES
  Vol 1: WALES
THE RIDGES OF SNOWDONIA
SCRAMBLES IN SNOWDONIA
SEVERN WALKS
THE SHROPSHIRE HILLS
SNOWDONIA WHITE WATER, SEA AND
  SURF
SPIRIT PATHS OF WALES
WELSH WINTER CLIMBS THE MIDLANDS
CANAL WALKS Vol: 2 Midlands
THE COTSWOLD WAY

COTSWOLD WALKS (3 volumes)
THE GRAND UNION CANAL WALK
AN OXBRIDGE WALK
WALKING IN OXFORDSHIRE
WALKING IN WARWICKSHIRE
WALKING IN WORCESTERSHIRE
WEST MIDLANDS ROCK

SCOTLAND
WALKING IN THE ISLE OF ARRAN
THE BORDER COUNTRY - A Walker's Guide
BORDER PUBS AND INNS - A Walker's Guide
THE CENTRAL HIGHLANDS - 6 LONG
  DISTANCE WALKS
CAIRNGORMS - WINTER CLIMBS
WALKING THE GALLOWAY HILLS
WALKING IN THE HEBRIDES
WALKS IN THE LAMMERMUIRS
WALKING IN THE LOWTHER HILLS
NORTH TO THE CAPE
THE ISLAND OF RHUM
THE ISLE OF SKYE - A Walker's Guide
THE SCOTTISH GLENS
  Book 1: CAIRNGORM GLENS
  Book 2: THE ATHOLL GLENS
  Book 3: THE GLENS OF RANNOCH
  Book 4: THE GLENS OF TROSSACH
  Book 5: THE GLENS OF ARGYLL
  Book 6: THE GREAT GLEN
  Book 7: THE ANGUS GLENS
  Book 8: KNOYDART TO MORVERN
  Book 9: THE GLENS OF ROSS-SHIRE
SCOTTISH RAILWAY WALKS
SCRAMBLES IN LOCHABER
SCRAMBLES IN SKYE
SKI TOURING IN SCOTLAND
TORRIDON - A Walker's Guide
WALKS FROM THE WEST HIGHLAND
  RAILWAY
THE WEST HIGHLAND WAY
WINTER CLIMBS BEN NEVIS AND GLENCOE

IRELAND
THE MOUNTAINS OF IRELAND
THE IRISH COAST TO COAST WALK
IRISH COASTAL WALKS

WALKING AND TREKKING IN THE ALPS
WALKING IN THE ALPS
100 HUT WALKS IN THE ALPS

CICERONE GUIDES

CHAMONIX TO ZERMATT - The Walker's
    Haute Route
THE GRAND TOUR OF MONTE ROSA
Vol 1: - MARTIGNY TO VALLE DELLA SESIA
    (via the Italian valleys)
Vol 2: - VALLE DELLA SESIA TO MARTIGNY
    (via the Swiss valleys)
TOUR OF MONT BLANC

FRANCE, BELGIUM AND LUXEMBOURG
WALKING IN THE ARDENNES
SELECTED ROCK CLIMBS IN BELGIUM AND
    LUXEMBOURG
THE BRITTANY COASTAL PATH
CHAMONIX - MONT BLANC
THE CORSICAN HIGH LEVEL ROUTE -
    Walking the GR20
WALKING THE FRENCH ALPS: GR5
WALKING THE FRENCH GORGES
FRENCH ROCK
WALKING IN THE HAUTE SAVOIE
TOUR OF THE OISANS: GR54
WALKING IN PROVENCE
WALKING IN THE LANGUEDOC
THE PYRENEAN TRAIL: GR10
WALKS AND CLIMBS IN THE PYRENEES
THE TOUR OF THE QUEYRAS
THE ROBERT LOUIS STEVENSON TRAIL
ROCK CLIMBS IN THE PYRENEES
WALKING IN THE TARENTAISE AND
    BEAUFORTAIN ALPS
ROCK CLIMBS IN THE VERDON
TOUR OF THE VANOISE
WALKS IN VOLCANO COUNTRY
THE WAY OF ST JAMES

FRANCE/SPAIN
ROCK CLIMBS IN THE PYRENEES
WALKS AND CLIMBS IN THE PYRENEES
THE WAY OF ST JAMES: Le Puy to Santiago
    - A Cyclist's Guide
THE WAY OF ST JAMES: Le Puy to Santiago
    - A Walker's Guide

SPAIN AND PORTUGAL
WALKING IN THE ALGARVE
ANDALUSIAN ROCK CLIMBS
COSTA BLANCA ROCK
MOUNTAIN WALKS ON THE COSTA BLANCA
ROCK CLIMBS IN MAJORCA, IBIZA AND
    TENERIFE
WALKING IN MALLORCA
BIRDWATCHING IN MALLORCA
THE MOUNTAINS OF CENTRAL SPAIN
ROCK CLIMBS IN THE PYRENEES
THROUGH THE SPANISH PYRENEES: GR11
WALKING IN THE SIERRA NEVADA
WALKS AND CLIMBS IN THE PICOS DE
    EUROPA

SWITZERLAND - including parts of France and Italy
ALPINE PASS ROUTE
THE BERNESE ALPS
CENTRAL SWITZERLAND
WALKS IN THE ENGADINE
THE JURA: WALKING THE HIGH ROUTE &
    WINTER SKI TRAVERSES
WALKING IN TICINO, SWITZERLAND
THE VALAIS, SWITZERLAND

GERMANY AND EASTERN EUROPE
GERMANY'S ROMANTIC ROAD A Guide for
    Walkers and Cyclists
WALKING IN THE HARZ MOUNTAINS
KING LUDWIG WAY
KLETTERSTEIG - Scrambles in the Northern
    Limestone Alps
THE MOUNTAINS OF ROMANIA
WALKING THE RIVER RHINE TRAIL
WALKING IN THE SALZKAMMERGUT
HUT TO HUT IN THE STUBAI ALPS
THE HIGH TATRAS

SCANDINAVIA
WALKING IN NORWAY

ITALY AND SLOVENIA
ALTA VIA - HIGH LEVEL WALKS IN THE
    DOLOMITES
THE CENTRAL APENNINES OF ITALY -
    Walks, Scrambles and Climbs
WALKING IN THE CENTRAL ITALIAN ALPS
WALKING IN THE DOLOMITES
WALKING IN ITALY'S GRAN PARADISO
LONG DISTANCE WALKS IN THE GRAN
    PARADISO
THE GRAND TOUR OF MONTE ROSA
ITALIAN ROCK - Selected Climbs in Northern
    Italy

CICERONE GUIDES

WALKS IN THE JULIAN ALPS
WALKING IN TUSCANY
WALKING IN SICILY
VIA FERRATA SCRAMBLES IN THE
   DOLOMITES

OTHER MEDITERRANEAN COUNTRIES
THE ATLAS MOUNTAINS
WALKING IN CYPRUS
THE MOUNTAINS OF GREECE
CRETE - THE WHITE MOUNTAINS
THE MOUNTAINS OF TURKEY
TREKS AND CLIMBS IN WADI RUM,
   JORDAN
JORDAN - Walks, Treks, Caves, Climbs,
THE ALA DAG, Climbs and Treks in Turkey's
   Crimson Mountains

THE HIMALAYAS
ADVENTURE TREKS IN NEPAL
ANNAPURNA - A Trekker's Guide
EVEREST - A Trekker's Guide
GARHWAL AND KUMAON - A Trekker's and
   Visitor's Guide
KANGCHENJUNGA - A Trekker's Guide
MANASLU - A Trekker's Guide
LANGTANG, GOSAINKUND & HELAMBU - A
   Trekker's Guide

OTHER COUNTRIES
MOUNTAIN WALKING IN AFRICA 1: KENYA

OZ ROCK - A Rock Climber's Guide to
   Australian Crags
TREKKING IN THE CAUCAUSUS
ROCK CLIMBING IN HONG KONG
THE GRAND CANYON AND THE AMERICAN
   SOUTH-WEST
ADVENTURE TREKS WESTERN NORTH
   AMERICA
CLASSIC TRAMPS IN NEW ZEALAND

CARTOONS - ideal gifts
LAUGHS ALONG THE PENNINE WAY
TAKE A HIKE
ON FOOT AND FINGER
ON MORE FEET AND FINGERS
THE WALKERS

CLIMBING, HILLWALKING AND TREKKING -
   Techniques and Education
ROPE TECHNIQUES
SNOW AND ICE TECHNIQUES
THE TREKKER'S MANUAL
THE HILLWALKER'S MANUAL
THE ADVENTURE ALTERNATIVE
FAR HORIZONS – Adventure Travel for all
MOUNTAIN WEATHER

EXPLORE THE WORLD
WITH A CICERONE GUIDE

Cicerone publishes over 280 guides for walking, trekking, climbing and
exploring the UK, Europe and worldwide. Cicerone guides are available
from outdoor shops, quality book stores and from the publisher. Cicerone
can be contacted on:

www.cicerone.co.uk

or at

2 Police Square, Milnthorpe, Cumbria LA7 7PY.
Please call for a full catalogue.

mountain / sports incorporating 'Mountain INFO'

Britain's liveliest and most authorative magazine for mountaineers, climbers and ambitious hillwalkers. Gives news and commentary from the UK and worldwide, backed up by exciting features and superb colour photography.

OFFICIAL MAGAZINE

Have you read it yet?

Available monthly from your newsagent or specialist gear shop.

Call 01533 460722 for details

BRITISH
MOUNTAINEERING
COUNCIL

THE WALKERS' MAGAZINE

THE GREAT OUTDOORS

**COMPULSIVE MONTHLY READING FOR
ANYONE INTERESTED IN WALKING**

*AVAILABLE FROM NEWSAGENTS,
OUTDOOR EQUIPMENT SHOPS, OR BY SUBSCRIPTION
(6-12 MONTHS) from*

**CALEDONIAN MAGAZINES LTD,
6th FLOOR, 195 ALBION STREET, GLASGOW G1 1QQ
Tel: 0141 302 7700 Fax: 0141 302 7799
ISDN No: 0141 302 7792   e-mail: info@calmags.co.uk**

Get ready for take off

Adventure Travel helps you to go outdoors over there. More ideas, information, advice and entertaining features on overseas trekking, walking and backpacking than any other magazine - guaranteed. Available from good newsagents or by subscription - 6 issues £15

Adventure Travel Magazine T:01789-488166

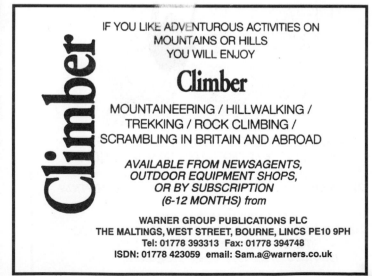

IF YOU LIKE ADVENTUROUS ACTIVITIES ON
MOUNTAINS OR HILLS
YOU WILL ENJOY

Climber

MOUNTAINEERING / HILLWALKING /
TREKKING / ROCK CLIMBING /
SCRAMBLING IN BRITAIN AND ABROAD

*AVAILABLE FROM NEWSAGENTS,
OUTDOOR EQUIPMENT SHOPS,
OR BY SUBSCRIPTION
(6-12 MONTHS) from*

WARNER GROUP PUBLICATIONS PLC
THE MALTINGS, WEST STREET, BOURNE, LINCS PE10 9PH
Tel: 01778 393313 Fax: 01778 394748
ISDN: 01778 423059 email: Sam.a@warners.co.uk